Vibe Coding Mastery

The Complete 5-in-1 Guide to Rapid AI-Powered Prototyping, Creative Dev Workflows, Code by Conversation, Low-Code Empowerment, and Next-Gen Explorer Mindset

Genne Yegge

GRAPEVINE BOOKS

Published by
GRAPEVINE BOOKS

www.grapevinebooks.com
email: contact@grapevinebooks.com

Ordering Information:
Quantity sales: Special discounts are available on quantity purchases
by corporations, associations, and others.
For details, reach out to the publisher.

First published by Grapevine Books, 2025

Contents

Preface: The New Developer Superpower

Software development has always been a moving target. Languages evolve. Frameworks appear, dominate, and fade. Tools that once felt revolutionary get replaced overnight. For decades, developers have had to run just to stay in place.

Then something new arrived. Not just a faster framework or a better library. Something that rewrote the rules entirely. Artificial intelligence moved into the heart of the development process. Suddenly, describing what you want could be as powerful as knowing how to code it. Tasks that took days could now happen in minutes. Problems that once felt too complex for a single developer could be tackled alone.

This is not hype. It is not theory. It is a daily reality for developers who choose to work with AI instead of around it. The results are staggering: prototypes in hours, complete apps in days, and the ability to explore unfamiliar tech without slowing down. This is the heart of vibe coding.

Vibe coding is not just about speed. It is about working in a way that feels natural and creative. Instead of grinding through syntax, setup, and boilerplate, you can focus on the flow of ideas. You explain your intent to an AI assistant, refine what it gives you, and keep moving. Every interaction feels like a conversation between your vision and the machine's ability to execute.

For new developers, vibe coding removes much of the initial friction. You can build something real in your first week of learning. You can experiment with concepts that might otherwise take months to understand fully. For experienced developers, vibe coding multiplies your reach. You can build features across languages and frameworks you have never touched before. You can move from concept to proof without waiting for team bandwidth or long release cycles.

For technical leaders, vibe coding changes how teams plan and deliver. Workflows can be redesigned around rapid iteration. Prototypes can be tested with real users before any full-scale build. The line between "research" and "production" becomes more flexible, allowing innovation without sacrificing quality.

But there is another layer to this shift. Once you start working in this new way, your mindset changes. You stop seeing coding as an endless list of technical hurdles and start seeing it as a creative space with fewer limits. That mindset shift is where the real power lies. AI becomes more than a tool. It becomes a partner that helps you explore, experiment, and push boundaries.

This book is a field guide to that world. It is built for developers, makers, and leaders who want to move fast without losing control of quality. It is for those who want to use AI as a creative amplifier, not just a code generator. It will give you frameworks, workflows, and examples you can use immediately. It will also help you develop the explorer mindset that will keep you adaptable as AI tools evolve.

By the end, you will not only know how to build faster. You will know how to think differently about what building means. The shift from syntax to intent is not a small one. It changes the role of the developer, the scope of what can be built, and the joy of the work itself. This is the moment to step into that shift with purpose. The future is already here. Let's make it yours.

Vibe Coding at a Glance: The Core Framework

Before we dive into techniques, tools, and case studies, it is worth stepping back to look at what vibe coding actually means in practice. Think of this as your mental map for the rest of the book. Every chapter you read will fit into one or more of these five pillars. Together, they form the complete framework for working at the speed and creativity AI makes possible.

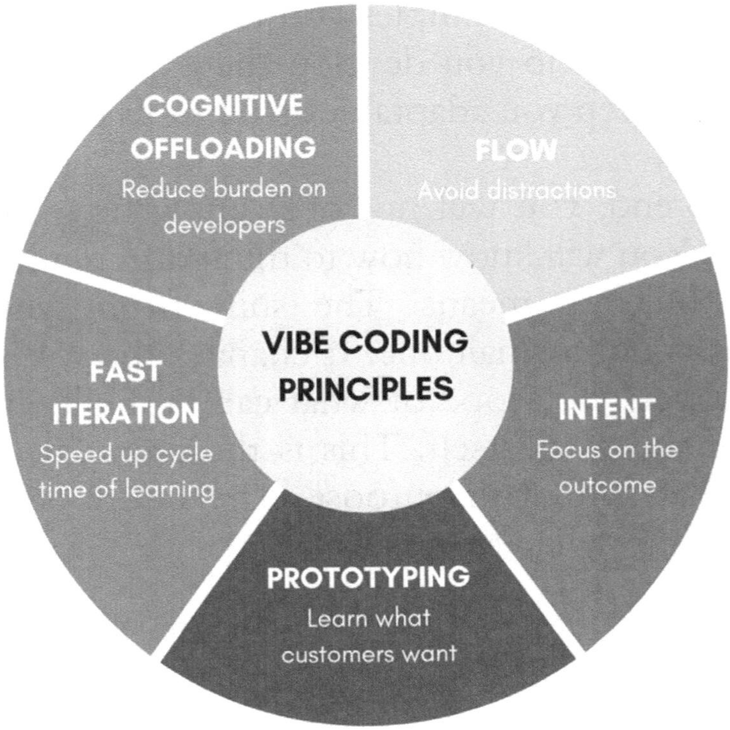

DEFINITIVE PRINCIPLES OF VIBE CODING

1. Rapid AI-Powered Prototyping

This is the foundation of vibe coding. It is the skill of turning ideas into working demos faster than most people can write a design doc. With the right AI-powered workflow, you can go from concept to clickable prototype in hours, sometimes minutes.

The emphasis here is on momentum. You do not wait to gather all the details or plan every feature. You focus on the one or two core actions that will bring the idea to life, and you use AI to fill in the gaps.

In vibe coding, prototypes are not throwaway sketches. They are functional, testable, and ready to evolve into real products. This speed changes how you think about building. You no longer need to choose between "thinking big" and "starting small." You can do both.

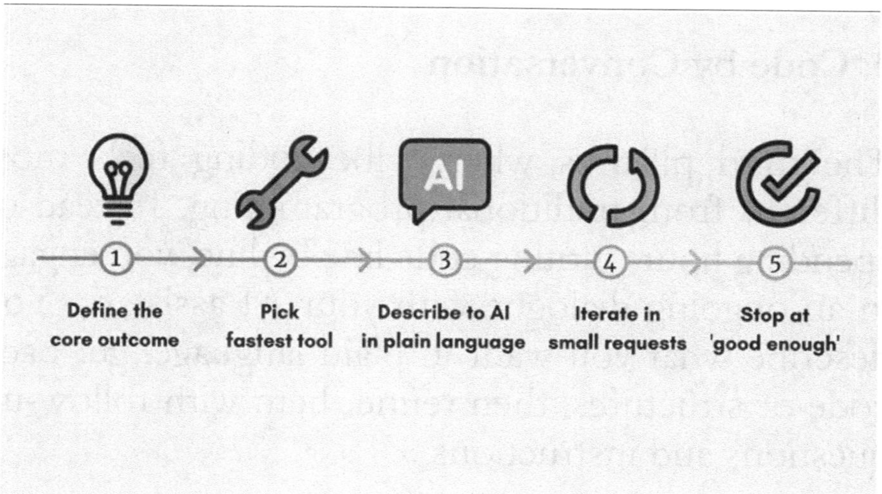

2. Creative Development Workflows

Speed without creativity leads to forgettable work. Creativity without structure leads to chaos. Creative development workflows bring both together. AI becomes your collaborator, helping you try variations, explore alternative designs, and quickly test ideas that would be too slow to explore manually.

This pillar is about designing your process so that creativity is not a lucky accident, but a repeatable part of your work. You will learn how to feed AI prompts that unlock unexpected solutions, how to remix existing codebases into new shapes, and how to create an environment where your best ideas have space to emerge.

3. Code by Conversation

The third pillar is where vibe coding feels most different from traditional programming. Instead of spending hours writing code line by line, you engage in an ongoing dialogue with your AI assistant. You describe what you want in plain language, get back code or structures, then refine them with follow-up questions and instructions.

The more skilled you get at this conversational style, the more complex and high-quality your output becomes. This is not about handing all the work to AI and hoping for the best. It is about steering the AI with clarity, precision, and an understanding of how to layer requests so you get exactly what you need.

4. Low-Code Empowerment

Vibe coding is not limited to traditional code editors. Low-code platforms can be a massive accelerator when combined with AI assistance. They allow you to integrate databases, APIs, automation tools, and interfaces without spending days wrestling with syntax.

In this pillar, the goal is to make low-code work for you, not box you in. You will see how to combine AI-generated code with low-code systems to add custom features, when to push beyond platform limits, and how to build solutions that are fast to launch but still flexible enough to grow.

5. Next-Gen Explorer Mindset

Finally, vibe coding is not just a set of tools or techniques. It is a mindset. Technology is evolving too quickly for static skill sets to remain valuable for long. The most important skill you can develop is the ability to explore new tools, frameworks, and ideas with confidence.

The explorer mindset means you can pick up a language you have never used before and start building with it in an afternoon. It means you can test new AI models and platforms without fear of "wasting time," because every experiment builds your adaptability. It means you can spot emerging trends and know when to pivot.

In vibe coding, you do not wait for the future to arrive. You walk toward it.

How the Framework Works Together

These five pillars are not separate silos. They reinforce each other. Rapid prototyping helps you generate and test creative ideas. Creative workflows make your prototypes more engaging and unique. Code by conversation speeds up your learning in both traditional and low-code environments. Low-code empowerment lets you deploy faster and reach more people. The explorer mindset ties it all together, ensuring you can keep adapting no matter what changes next.

By the time you have mastered all five, you will not just be coding faster. You will be creating at a scale and quality that would have been impossible a few years ago. And you will be ready for whatever comes next in this AI-powered world.

Book 1: Rapid AI-Powered Prototyping

Introduction: Why Speed is the First Superpower

Speed is not just a convenience in modern development. It is a competitive weapon. In an era where new products and features appear daily, the ability to move from idea to working prototype in hours can be the difference between leading a trend and chasing it. Rapid prototyping is not only about making something quickly. It is about compressing the distance between your first spark of inspiration and a tangible, testable result that you can put in front of users, stakeholders, or even yourself.

In the past, prototyping often meant long setup times and slow iteration cycles. You might spend a full day just installing dependencies, configuring environments, and preparing boilerplate before you could write the first real line of code. Then came the coding itself, a process that could take weeks before you had something that resembled a usable product. By the time you got there, market conditions might have changed, user needs might have evolved, and your original excitement could have faded.

With vibe coding, this reality is turned inside out. AI-assisted development eliminates many of the friction points that used to slow us down. You do not need to memorize every syntax detail. You do not have to navigate endless documentation before you can take a first step. You can describe what you want, get working code in seconds, and refine it in a conversational loop that feels fluid and fast.

The impact of this speed is profound. You can validate ideas before committing significant time or resources. You can test multiple variations of a concept in a single day. You can give early prototypes to real users and watch how they respond, which is infinitely more valuable than guessing how they might respond.

Rapid prototyping also changes the emotional tone of development. Instead of being weighed down by technical hurdles in the early stages, you ride a wave of momentum from the moment you begin. You see results quickly, which fuels your motivation to keep going. The cycle becomes self-sustaining: fast progress leads to more experimentation, which leads to better ideas, which leads to even faster progress.

In this part of the book, we will break down the specific skills and workflows that make rapid AI-powered prototyping not only possible but repeatable. You will learn how to define the core of an idea so you can build it quickly. You will learn how to choose the right tools for speed without sacrificing too much flexibility. You will learn how to enter completely new technical territory and still ship a working prototype the same day. And you will learn how to use small, focused micro-projects to sharpen your skills while producing something useful.

By the time you finish this section, you will have a blueprint for turning almost any idea into a functioning, testable product in hours instead of weeks. More importantly, you will have the confidence that you can do it again and again, no matter what technology, framework, or toolset is involved.

Chapter 1: From Idea to Demo Before Lunch

The biggest shock for anyone stepping into AI-assisted development is just how quickly you can go from a blank screen to something real. In the past, building even a simple app meant carving out days, sometimes weeks, before you could show anything that actually worked. You had to research frameworks, install dependencies, set up environments, plan your architecture, write the first version of your code, debug it, and only then would you have a product worth demonstrating. That slow grind often meant ideas died before they even had a chance to be tested. With AI in your corner, that timeline collapses. You can walk into your morning with nothing more than a half-formed idea in your head, and by lunchtime you can have a working prototype on your screen, ready to show to others.

The key shift is to stop thinking of a prototype as a polished product. In vibe coding, a prototype is not a "draft" in the way people often mean it. It is a living, working model of your idea, stripped down to the essentials. It might look rough. It might be missing features. It might not even be fully secure or optimized. But it works. It shows the magic of the idea in action, and that is what matters at this stage. Your goal is not to impress with perfection, but to inspire with possibility. Once you get something working quickly, you can refine and expand it based on real feedback rather than assumptions.

To make this happen, you need a mental filter for scope. Ask yourself: what is the absolute smallest version of my idea that still delivers the "aha" moment? If you want to build a language-learning platform, maybe the core is not a complete user account system with leaderboards and lesson tracking. Maybe the core is simply a single web page that takes a phrase in one language and returns a translation, along with a simple quiz to reinforce it. If you want to build a productivity app, perhaps you skip the mobile version for now and just make a web dashboard that demonstrates the key interaction. AI thrives in this kind of focused environment, because you can describe a narrow, clear goal and get something back fast.

Once you have defined that smallest core, you can begin the process. The most efficient way to get from idea to demo is to break the build into three short phases. The first is framing, where you describe your idea clearly to your AI assistant in natural language. Do not start with code right away. Write it out as if you were explaining it to a teammate who has never heard of your project before. State the purpose, the primary function, and any specific technologies you want to use. For example: "I want to make a simple HTML page that lets the user paste text, click a button, and get a two-sentence summary from an AI model. The page should look clean and modern, and it should work without the user needing to log in." This gives the AI a complete picture of the task before it starts producing output.

The second phase is layered building. Do not ask for the entire application in one request. Break it into components you can test and refine quickly. Start with the HTML skeleton. Once that works, ask the AI to generate the JavaScript that connects it to your AI backend. Once that is functional, move to styling. With each step, you are making progress you can see and touch, and you can correct mistakes before they compound. This iterative loop is where AI shines, because it never gets tired of your follow-up requests. It will adjust the code as many times as you need, without frustration or delay.

The third phase is polishing for presentation. Even though you are working fast, it is worth spending a small block of time making the prototype look and feel a bit better before you show it. This is not about adding features. It is about making sure the experience communicates your idea clearly. Tweak the text labels so they are human-friendly. Align elements on the screen so they look intentional. Give the interface just enough design so that people can focus on the concept rather than the roughness of the build. This small investment pays off, because a visually clear prototype is far more persuasive when you present it.

Here is an example to make it concrete. Imagine you wake up with the idea for a "Quick Mood Journal" app that helps people capture their feelings and get a short reflection or tip from an AI. At **9:00** AM, you sit down and tell your AI assistant: "Make me a single web page with a text area for the user to type how they feel. Add a submit button. When clicked, send the text to an AI model and return a short, empathetic reflection." By **9:10**, you have a basic HTML form. You then ask the AI to connect it to an API endpoint that can call an AI model. By **10:00**, the connection works. By **11:00**, you have added some CSS for a calming look and tested it a few times to make sure the responses feel friendly. By lunch, you have a working prototype you can share with friends or colleagues for feedback. Without AI, this would have taken you days.

The real magic is not just the speed. It is what the speed makes possible. When you can create working prototypes in a morning, you stop being cautious about exploring ideas. You try more things. You get more feedback. You discover new directions you could not have predicted at the start. And each small success fuels your confidence and skill, which makes the next idea even easier to bring to life. This is the engine of rapid AI-powered prototyping, and it is the reason vibe coding feels like a superpower.

Chapter 2: Selecting the Fastest AI Stack for the Job

In vibe coding, speed is not about cutting corners. It is about removing the unnecessary friction between your idea and its first working form. And one of the most important factors in determining that speed is your choice of stack - the combination of tools, languages, frameworks, platforms, and services that you use to bring your prototype to life.

The right stack can feel like a launch ramp, propelling you forward with minimal resistance. The wrong stack can feel like running through mud, where even small steps are slow and exhausting. In traditional development, stack selection was often based on long-term maintainability, deep familiarity, or what the team already used. In rapid AI-powered prototyping, the selection criteria are different. You are not committing to a lifetime relationship with these tools. You are picking the ones that will get you to a working demo in the shortest time possible, without creating so many compromises that your prototype loses credibility.

Understanding What "Fast" Really Means

Speed in prototyping is not measured in raw performance benchmarks or in how quickly a framework can compile. Speed is measured in time-to-first-result; how long it takes you to see your idea doing something real on the screen. That first visible, testable output is the moment you switch from imagining to interacting. Once you can interact with your idea, every improvement after that is easier to make because you have something tangible to guide you.

When you evaluate a stack for speed, think about:

- **Setup Time:** How long before you can write your first feature?
- **Learning Curve:** Can you be productive without reading pages of documentation?
- **Integration Friction:** How easy is it to connect with APIs, databases, and AI services?
- **Deployment Speed:** Can you share your prototype with someone else in minutes, not days?

These factors will matter more than whether the framework is "industry standard" or has a perfect long-term architecture. Remember, a prototype is meant to be a working experiment, not the final cathedral.

AI as a Stack Multiplier

AI changes stack selection because it reduces the penalty for using something unfamiliar. In the past, a developer might avoid a language they did not know well because the learning curve would be too steep. Now, you can pick up an unfamiliar stack and have the AI fill in the gaps while you build. You can say, "Generate a simple CRUD app in SvelteKit that connects to a Supabase database" and watch a working draft appear in seconds. You do not have to know every command or syntax pattern in advance - the AI can generate, explain, and adapt as you go.

This opens up possibilities. You can choose the stack that is fastest for the job, not just the stack you already know. You can step outside your comfort zone without losing time. And over time, your comfort zone itself expands, giving you even more flexibility for future builds.

Choosing the Fastest Stack for the Job

Here is the process I recommend when deciding on a stack for a rapid AI-powered prototype.

Step 1: Identify the Output First

Before you even think about languages or frameworks, decide exactly what you want to produce. Is it a web app, a mobile app, a data processing script, an automation workflow, or an embedded widget? The desired output will narrow your options immediately. For example:

- A quick interactive web interface? Consider HTML/CSS/Vanilla JS, or a lightweight framework like Svelte or Next.js.

- A mobile MVP? FlutterFlow or React Native with AI-assisted code generation.

- Data processing or analytics? Python with Pandas, Streamlit for quick dashboards.

- Internal business tool? Retool or WeWeb combined with AI extensions.

Step 2: Choose for Instant Visibility

Pick tools that let you see results almost immediately. For instance, if your prototype needs a UI, choose a framework or platform with hot reload or instant preview. If it is backend-only, choose one that lets you test endpoints with minimal setup, like FastAPI or Flask in Python.

Step 3: Use Platforms With Built-in Deployment

Being able to deploy instantly is critical for feedback loops. Platforms like Vercel, Netlify, Railway, or Replit let you go from local code to a live link in minutes. This is invaluable when you want to show someone your work before the end of the day.

Step 4: Leverage Low-Code Where It Fits

You do not have to pick between low-code and full-code, you can mix them. A low-code UI layer with AI-generated backend logic can be far faster than building everything from scratch, especially for prototypes.

The "Speed Stack" Concept

Every developer should maintain a personal Speed Stack — a shortlist of tool combinations you know you can set up and run quickly for specific kinds of projects. This is not your "ideal stack" for production; it is your go-to kit for prototyping. Here is an example:

- **Web App MVP**: Next.js + Tailwind CSS + Supabase + OpenAI API, deployed on Vercel.
- **Data Dashboard:** Streamlit + Pandas + Plotly +

GPT API for summaries.

- **Mobile MVP:** FlutterFlow with AI-generated Dart snippets.
- **Automation Tool:** n8n or Make.com with custom AI integration nodes.
- **API Prototype:** FastAPI + SQLite + GPT API, deployed on Railway.

When you have a Speed Stack ready, you are never starting from zero. You know exactly what to reach for, and you can tell your AI assistant the entire setup in one sentence.

Balancing Familiarity and Capability

The temptation with AI is to always try something new, because the learning curve is less painful. But there is a trade-off: unfamiliar tools may still introduce delays if they have quirks you do not expect. The fastest stack is often the one you already know that is good enough for the prototype, even if it is not "perfect" for the final product.

The right balance is to lean on your familiar tools for time-sensitive builds and deliberately explore new stacks when you have a little more breathing room. That way, your Speed Stack grows without slowing your current project.

Deployment as Part of the Stack

One of the most overlooked aspects of stack selection is deployment. A prototype that works locally but cannot be easily shared is far less valuable. If it takes hours to get a live demo running, you lose momentum and delay feedback.

When evaluating a stack, ask:

- Does it have one-click or command-line deployment to a cloud host?
- Can I get a live link without messing with DNS or server configs?
- Will it handle basic scaling if a few dozen people try it at once?

A stack that includes a deployment-ready host like Vercel, Netlify, or Replit can save you a day or more of setup.

Example: The Fastest Path to a Working App

Imagine you are tasked with building a simple tool that lets users paste a block of text, have it summarized by AI, and download the summary as a PDF. With the wrong stack, you could lose hours before you even see a basic interface. With a smart choice, you can be done before lunch.

Option 1: Build in React with a custom backend, set up your own API endpoints, manage deployment manually. This could take hours just for setup.

Option 2: Use Streamlit for the UI, Python's requests library for the API call, and fpdf for PDF creation. Deploy to Streamlit Cloud. You can go from zero to a working, sharable app in under **90** minutes.

The difference is not in how "good" the stack is in the abstract. It is in how quickly it moves you from idea to live demo.

The AI Edge in Stack Assembly

AI does not just help you code within a stack, it can help you choose and configure the stack itself. You can ask:

- "Given my goal to build [X], what is the fastest stack I can use that will deploy today?"
- "Generate a setup script to create a new project in [framework], add [library], and connect to [API]."
- "List the pros and cons of building this prototype

in [Tool A] vs [Tool B] if my deadline is tomorrow."

The AI can even produce boilerplate code and configuration files, so you skip the manual steps and start with a working skeleton.

The takeaway is simple: in rapid AI-powered prototyping, the right stack is the one that gets you to interaction the fastest. You are not making a long-term bet here; you are picking the shortest bridge from idea to reality. The faster that bridge takes you across, the more ideas you can test, and the more often you can win.

Chapter 3: Prototyping in New Languages and Frameworks Instantly

For many developers, the decision to work in a new programming language or framework has historically been a big one. It meant stepping into a steep learning curve, weeks of tutorials, and a lot of trial and error before you could produce something worth showing. The cost (in time, effort, and sometimes frustration) made it tempting to stick to what you knew, even when another tool might have been a better fit for the job.

Vibe coding changes that equation completely. With AI as your partner, the barrier to entry for new technologies is lower than it has ever been. You no longer need to master a language before you can use it effectively. You can start building with it immediately, using the AI to generate, explain, and refine code in real time. Instead of "learning first, building later," you can learn while building — and ship a working prototype in the same session you first touched the technology.

Why This Is a Game Changer

Before AI-assisted workflows, switching to a new language or framework often required:

- Reading through official documentation front to

back.

- Searching for working examples in blog posts or GitHub repos.
- Piecing together bits of knowledge from forums.
- Spending hours debugging basic setup issues.

Even if you already had strong development skills, this onboarding period could be long enough to kill momentum. Now, with conversational AI, you can cut through that entire pre-build phase. You can go from zero knowledge to a running prototype in hours because the AI handles the translation between your intent and the new stack's syntax and conventions.

The psychological impact of this is just as important as the technical one. When you know you can explore a new technology without a massive time penalty, you become more adventurous. You stop asking, "Do I have time to learn this?" and start asking, "What could I make if I tried this today?"

The Mindset for Rapid Entry

The key to prototyping in new languages instantly is to adopt the mindset of a builder first, learner second. That might sound backward compared to traditional programming advice, but it works because of how

AI shifts the learning process.

You are not trying to absorb an entire language reference up front. You are aiming to produce a working piece of software and letting the AI handle the parts you do not yet know. In the process, you will naturally learn the essentials because you are using them in context. This is active, purpose-driven learning, it sticks better because it is tied to something you care about building.

The "Minimal, Expand, Refine" Process

When entering an unfamiliar tech stack, this three-step process keeps you moving quickly without losing control.

Step 1: Minimal

Ask the AI for the simplest possible implementation of what you want in the new language or framework. For example:

"Show me the minimal code to build a single-page app in Vue.js that displays a list of items from an array."

Run this minimal example as soon as possible. Seeing it work confirms your environment is set up correctly and gives you a base to build on.

Step 2: Expand

Once the minimal version is running, ask for one new feature at a time. Keep it small. Add a button. Connect to an API. Add a bit of styling. Test each change before moving on. The AI can generate these additions in seconds, but you control the pace so you can understand what is happening.

Step 3: Refine

Once you have the core features, clean them up. Ask the AI to organize the code into standard project structure for the framework. Request explanations for any unfamiliar patterns. This is where your quick prototype starts looking like something you could share proudly.

Example: First-Time Build in Go

Let's walk through a realistic example. Imagine you have never written a line of Go, but you want to build a simple REST API that returns a random motivational quote.

Framing the goal for AI:

"Build a minimal Go server with one endpoint / quote that returns a random motivational quote in JSON format."

The AI gives you a short Go program using net/http, a slice of quotes, and a random selection function. You paste it into your Go environment, run it, and see it working in the browser within **10** minutes.

Expanding:Next, you ask:

“Modify this so it reads quotes from a JSON file instead of hardcoding them.”

The AI adjusts the code to load the file at startup and return random entries.

Refining: Finally, you say:

“Restructure this into a standard Go project layout with a main.go file, a /handlers folder for endpoint logic, and a /data folder for the quotes file.”

In under an hour, you have a working Go API with a clean structure, despite starting with zero Go experience.

Using AI as Your Translator

One of the most powerful ways to accelerate in a new stack is to use AI as a syntax and concept translator. If you know how to do something in one language, you can describe it and ask for the equivalent in another. For example:

"In JavaScript, I would use map() to transform an array. Show me the Go equivalent with an example."

This lets you leverage what you already know instead of starting from scratch.

Avoiding Common Pitfalls

Even with AI's help, there are traps to avoid when building in a new language:

1. Overloading the First Build

It is tempting to ask for a complex, multi-feature app right away. Resist this. Complexity compounds the risk of errors you cannot yet diagnose. Start simple, layer complexity.

2. Ignoring the Explanations

If you treat the AI purely as a code vending machine, you will miss the chance to actually learn the new stack. Take the time to ask why something was done a certain way.

3. Copy-Paste Without Testing

Run and test after every change. Do not let multiple untested modifications pile up, it makes debugging harder.

4. Skipping Environment Knowledge

While you do not need to master the language immediately, you should at least understand how to run, test, and package code in it. Ask the AI for these basics early.

Long-Term Benefits

Even if you never use that new language again for a full project, the experience builds your adaptability muscle. It makes you comfortable with jumping into unknown territory and finding your way quickly. Over time, this reduces the intimidation factor of emerging tools and frameworks.

You will also build a mental library of "entry points", patterns for starting projects in various environments. The next time someone mentions a new tool, you will be able to say, "I could spin up a quick demo in that today", and mean it.

The Confidence Shift

Perhaps the most important outcome of this approach is the confidence it gives you. When you know you can produce a working prototype in an unfamiliar language in the space of a single day, you stop being afraid of change. You stop clinging to the stacks you know just because they are familiar.

You become a technology explorer rather than a technology survivor.

In vibe coding, this ability to prototype in any stack is like having a passport that works everywhere. It makes you faster, more versatile, and far more resilient in an industry that never stops evolving. The tools will change. The languages will change. But your ability to adapt will stay with you.

Chapter 4: Micro-Projects That Teach You Skills While You Build

One of the most underrated ways to become an unstoppable vibe coder is to think small on purpose. Not small in ambition, but small in scope. The truth is, large projects can teach you a lot, but they also trap you in long timelines, complex dependencies, and sometimes months of work before you see any payoff. Micro-projects are different. They give you an immediate win, a tangible deliverable, and a focused learning outcome. You finish them in hours, not weeks, which means you get more cycles of experimentation and growth.

The beauty of micro-projects is that each one can be designed to teach you something specific while still producing something you can use, share, or even sell. Instead of sitting down to "learn a framework" in the abstract, you learn it by building a tiny app that forces you to touch the parts of the framework that matter most. Instead of trying to master a new API by reading docs for hours, you integrate it into a single, sharply defined feature that's up and running the same afternoon.

Why Micro-Projects Work So Well for Vibe Coding

Micro-projects are like perfectly sized playgrounds for AI-assisted development. Their compact scope allows you to work on something meaningful without feeling overwhelmed, and that scale matches the strengths of AI in a way larger projects can't always replicate. Because everything fits neatly into a single conversation, the AI can hold the entire project in context. It doesn't have to piece together disconnected fragments of your code from memory. It can "see" and reason about the whole thing at once. This makes its output more coherent, more consistent, and often more creative.

The result is that you and the AI can talk through the build like two collaborators standing at the same whiteboard. You can describe your goal, receive a complete version of the feature, test it immediately, and then refine it without losing the thread. There's no need to constantly re-explain the architecture or paste huge chunks of code to get the AI back up to speed. That kind of continuity is harder to achieve in sprawling, multi-file projects where context limits become a real barrier.

Micro-projects also shine because they make the feedback loop extremely short. In a larger system, even a small change can ripple through dozens of files, making it harder to predict what will break and harder to confirm you've fixed it. In a micro-project, there's less surface area to worry about. If something isn't working, you can track it down quickly, fix it, and immediately see the result. That immediacy builds a rhythm where every iteration feels like progress, and each small success reinforces your momentum.

This matters enormously in the early stages of mastering vibe coding. Momentum is fuel, and nothing drains it faster than getting bogged down in a wall of complexity you don't yet understand. When you're building something small, there are fewer places for problems to hide, which means you spend less time lost in the weeds and more time in flow. The AI stays in sync with you, you stay engaged with the build, and the learning feels natural because every change produces a visible, tangible effect.

That's why micro-projects don't just make you faster, they make you better. They give you a safe, contained space to experiment, learn new techniques, and develop confidence in your AI-assisted workflow, all without the weight of a massive codebase slowing you down.

How to Design a Micro-Project for Maximum Learning

The trick to making micro-projects valuable is to be intentional about their design. You want each one to be fun, fast, and focused; three qualities that guarantee you'll finish it and learn something along the way.

1.Pick One Clear Objective

Every micro-project should have a single primary goal. That might be to learn how to connect to a database, to practice conversational prompting for AI, or to try a new front-end framework. Avoid piling on unrelated features "just because you can." If the project's purpose is to display live weather data, resist the urge to also add user accounts, complex settings menus, or payment processing. Keep the focus sharp.

2.Set a Strict Time Limit

A true micro-project should be completable in a single sitting, ideally in under four hours. This forces you to make decisions quickly, avoid unnecessary detours, and lean on AI for speed. It also makes it easy to stack multiple micro-projects into your learning schedule without them taking over your life.

3.Embed a New Skill or Concept

Make sure the project forces you to use something you've never used before or in a way you haven't tried before. It might be a new language, a specific API, or a design pattern. Even if the output isn't revolutionary, the skill you gain will carry over into bigger projects later.

4.Make It Shareable

Even if the project is small, design it so you can show it to someone; a friend, a colleague, or the online developer community. That little push to present it will motivate you to polish it just enough to make it enjoyable to use or view.

5.Document As You Go

Instead of waiting until the end, write quick notes as you build. This can be as simple as a text file with bullet points like "Learned how to use fetch() in Vanilla JS" or "Supabase quick start was faster than I expected." These notes become a personal knowledge base you can revisit later.

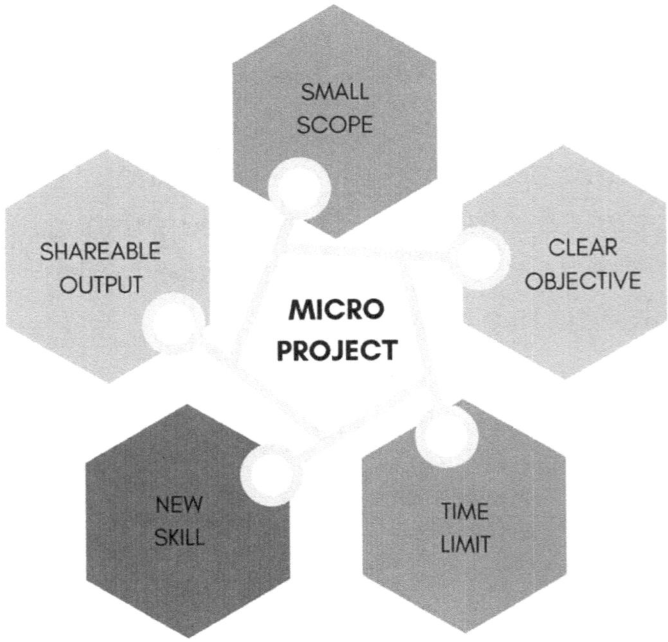

Examples of High-Value Micro-Projects

To give you a sense of what this looks like in practice, here are a few examples that fit perfectly into the vibe coding approach:

- **AI-Powered Text Summarizer**

A single web page where a user pastes text and gets a concise summary from an AI model. Teaches: API integration, prompt tuning, handling async requests.

- **Image-to-Color Palette Extractor**

Upload an image, display its dominant colors, and let the user copy hex codes. Teaches: file uploads, image processing APIs, UI color handling.

- **Simple Habit Tracker**

Lets users add a habit, mark it complete for the day, and view a **7**-day streak. Teaches: basic CRUD operations, local storage or lightweight database use.

- **Markdown-to-PDF Converter**

Paste markdown text, click a button, and download it as a styled PDF. Teaches: file handling, text processing libraries, frontend-to-backend interaction.

- **Live Crypto Price Ticker**

Pulls real-time prices from a public API and displays them in a minimal dashboard. Teaches: fetching data, formatting, and simple animations.

- **Quick Portfolio Site Generator**

Prompt the AI for portfolio content, drop it into a styled HTML template, and deploy it. Teaches: static site generation, deployment workflows.

How to Run a Micro-Project Session

When you sit down to do a micro-project, think of it as a focused, high-intensity workshop where you're both the student and the builder. You're there to learn, but you're also there to ship something, even if it's tiny. The aim is to create a container of time and attention where you can explore, build, and finish without getting pulled into distractions or endless polish.

Start with a completely clean slate. This isn't about picking up half-finished work or "finally getting around" to something. Open a new workspace, spin up a fresh project in your low-code tool, or create a new repo, whatever makes it feel like the start of a new experiment. Then set a timer. Ninety minutes is often ideal, but even a single hour can work. The time limit is important because it forces you to make decisions quickly instead of disappearing down rabbit holes.

The first thing you tell your AI is exactly what you want to build and why. The "why" matters because it helps the AI focus on the core purpose instead of loading you down with irrelevant extras. For example, "I want to build a basic weather dashboard for learning API integration" will guide the AI toward a clean, functional solution, whereas "Build me a weather dashboard" might invite more complexity than you need.

As the AI produces code or config, test it immediately. Don't let code sit untouched for more than **10** minutes without running it. Rapid feedback is the lifeblood of micro-projects. You want to see things work or fail, as soon as possible, so you can adjust your approach in real time. Every successful run is a small win; every failure is a fast lesson.

And don't shy away from the failures. In fact, expect them. Part of the point of a micro-project is to bump into problems while the stakes are low. When you hit an error, copy it directly into your AI chat and ask for two things: the fix and the explanation. Even if you don't fully understand the explanation in the moment, it plants a seed. Later, you'll encounter a similar issue and recognize it immediately, shortening the learning curve. Because the explanation is tied to something you just experienced, it will stick far better than any example you'd skim in a tutorial.

When the timer buzzes, stop, no matter how close you feel to "just one more tweak." Ship the project as it is. That might mean deploying it to a quick URL, showing it to a friend, or simply saving it in your micro-project archive. The point is to declare it done, take your notes, and move on.

Over time, the compounding effect of many small builds is far greater than the perfection of any single one. A dozen imperfect projects teach you more, and give you more tools, patterns, and confidence than one big project you agonize over for weeks. Micro-project sessions are about momentum, and momentum is the foundation for mastery in vibe coding.

The Compounding Power of Micro-Projects

The magic of micro-projects is that they build on each other in subtle but powerful ways. The first time you integrate an API, it might take you an hour to figure out authentication. The next time, it takes **10** minutes. The first time you deploy a site, you might wrestle with DNS settings. The next time, you breeze through it without thinking. Each project removes friction from your future work.

Over time, you start to see patterns. You realize that the same few steps get repeated in most builds, so you automate them. You notice which stacks give you the least trouble and which ones slow you down, so you adjust your defaults. You build a toolkit, not just of code snippets, but of instincts. And instincts are what separate a hesitant builder from a confident one.

Micro-Projects as a Creative Engine

There is also a creative dimension here. Large projects often lock you into one vision for months at a time. Micro-projects free you to explore wild ideas without risk. You can build a silly "Haiku Generator" in the morning and a functional "PDF Invoice Tool" in the afternoon. Some will remain as fun experiments, but others will spark bigger ideas worth pursuing seriously. Many startups and products have begun as tiny side experiments that unexpectedly resonated with people.

The beauty is that you never know which small build will open the next big door. All you have to do is keep making them.

When you combine rapid AI-powered prototyping with the discipline of micro-projects, your learning accelerates, your output multiplies, and your fear of new technology dissolves. You stop seeing the unknown as a roadblock and start seeing it as an invitation. And with every completed micro-project, you become a little more dangerous, in the best possible way, as a creator.

Chapter 5: Designing Prototypes That Invite Feedback Early

A prototype is a bridge between imagination and reality. It is not the finished building, but it lets people walk across, feel the structure, and tell you whether they want to go further. In vibe coding, because you can build prototypes so quickly, you have the unique advantage of making that bridge appear almost overnight. The question is: do you use it just to prove that you can cross it, or do you invite others to walk with you and tell you what they see?

Too many builders treat prototypes as private proofs of concept — personal experiments kept under wraps until they are "ready." This is a missed opportunity. A prototype becomes far more valuable when it is used as a conversation starter, a feedback magnet, and a way to explore possibilities through other people's eyes.

In this chapter, we will go deep into how to make prototypes that actively invite feedback, not just from developers, but from real users, potential stakeholders, and even people who might challenge your thinking in ways that make the final product stronger.

The Power of Early Feedback

The sooner you show a prototype, the sooner you discover whether you are building in the right direction. Feedback at the start is exponentially more valuable than feedback after weeks or months of development. This is because early changes are cheap; in time, money, and emotional investment.

Think of it like turning a small boat versus turning a giant ship. In the early days, your "boat" is light and nimble. You can pivot in minutes. Later, when you have a big codebase, a complex UI, and invested weeks of work, making a change can feel like moving mountains.

Early feedback also prevents founder's bias - the blind spot we all have when we fall in love with our own ideas. You may think a certain feature is the heart of your product, only to discover that users do not care about it. Or you may overlook a small detail that turns out to be the thing people love most.

What Makes a Prototype "Feedback-Friendly"

Not all prototypes are equally good at generating useful feedback. Some are too rough for people to understand. Others are too polished, giving the false impression that the product is finished and discouraging honest criticism.

The sweet spot is clear enough to communicate the idea but open enough to invite suggestions. That means:

1.**Focused scope** — Show only the main feature or features you want feedback on.

2.**No distractions** — Remove placeholder pages, broken buttons, or half-implemented features unless they are relevant to the feedback you want.

3.**Visible purpose** — The user should be able to tell what the prototype is meant to do within seconds.

4.**Low entry barrier** — No complicated sign-up, no multi-step setup, no steep learning curve.

5.**Responsive enough** — Speed matters. If your prototype lags, feedback will be about performance rather than functionality.

Designing With Feedback in Mind

Before you start building, decide exactly what you want to learn from feedback. If you are clear about the questions you are trying to answer, you can design your prototype to make those questions easy for users to engage with.

For example:

- If you want to know whether your navigation flow is intuitive, build enough pages to let people explore a task from start to finish.
- If you want to test the value of a feature, make that feature prominent and easy to find.
- If you want to know how users respond to AI-generated content, fill the prototype with example outputs so they can react to real samples.

The AI Boost in Feedback Prototyping

AI can help make your prototypes more feedback-friendly in several ways:

- **Generate sample data** so the interface looks real and users can respond to realistic scenarios.
- **Create guided tours** that highlight key features for first-time testers.
- **Write concise descriptions** for the landing page that make the purpose instantly clear.
- **Quickly re-style** elements to make the interface clean enough for easy use, even if design is not your strength.

Example prompt:

"Write a short, friendly description for a prototype web app that summarizes news articles into bullet points. The description should make the purpose instantly clear to a first-time visitor."

Making the Feedback Process Frictionless

The design of your prototype is only half the equation. The other half is how you present it to testers. You want the feedback process to be as easy and inviting as possible.

1. Choose the right audience for the stage

In the earliest stage, you might share with fellow developers or friends who can tolerate rough edges.

As it stabilizes, you share with target users who can give more realistic reactions.

2. Give clear instructions

Do not just say "Tell me what you think." Give specific actions: "Add three tasks to the list, then try marking one complete. Tell me how that felt."

3. Make it easy to respond

Embed a quick feedback form, create a Slack/ Discord thread, or even just provide a single "Send Feedback" button that opens an email pre-filled with the prototype link and a few guiding questions.

Scripts for Presenting a Prototype

Here are some short scripts you can use when sharing your prototype:

For casual testers:

"I built this quick prototype for [X]. It's super early. Could you try [specific action] and tell me if it felt easy or confusing?"

For target users:

"This tool is meant to [solve problem]. Could you try it for [short task] and let me know if it solves that problem for you?"

For stakeholders:

"This is a proof-of-concept showing how we could [deliver benefit]. It's not final, but I'd like your reaction to the concept and whether it feels worth pursuing."

The 24-Hour Feedback Loop

One of the most powerful habits you can build in vibe coding is the **24**-hour feedback loop:

- **Day 1 Morning** — Build the core version of the feature or product.
- **Day 1 Afternoon** — Share it with a few people for feedback.
- **Day 2 Morning** — Review feedback, make targeted improvements.
- **Day 2 Afternoon** — Share the improved version.

This rhythm keeps you from drifting too far without input and builds a steady momentum of improvement.

A Real Example: AI-Powered Recipe Finder

Let's say you have the idea for an AI-powered recipe finder that takes ingredients you have on hand and suggests recipes. Using AI-assisted prototyping, you build the MVP in four hours:

- Simple input box for ingredients.
- AI call to generate recipe suggestions.

- Minimal styling to make it readable.

That afternoon, you send it to ten friends with the request:

"Try entering whatever's in your fridge right now and tell me if the suggestions are appealing."

From their feedback, you learn:

- People want images for each recipe.
- Some ingredient names are being misinterpreted by the AI.
- The results page feels cramped on mobile.

By the next day, you have:

- Added free stock images from an API.
- Tweaked prompts to handle ambiguous ingredient names.
- Adjusted the layout for mobile screens.

You did not need to guess what to improve. The users told you exactly where to focus.

Avoiding the Feedback Trap

Early feedback is gold, but it can also be noisy. Some people will suggest features that conflict with your vision. Others will give vague comments that are hard to act on.

The key is to look for patterns:

- If multiple testers hit the same obstacle, fix it.
- If only one person mentions something, ask yourself if it aligns with your goals before acting.

Not all feedback is equal, and not all suggestions should be implemented.

Building Feedback Into Your Culture

The best builders make feedback part of their DNA. It is not something they do once or twice — it is a habit. Every prototype they make is shown to someone else within **24** hours. Every major change is validated before it becomes a sunk cost.

When you work this way, you stop fearing criticism. You start seeking it out. You see each comment as another piece of the puzzle that makes your product better.

The Payoff in Vibe Coding

Vibe coding makes feedback-friendly prototyping easier than ever:

- You can create something to show in hours.
- You can adjust based on feedback in the same day.
- You can repeat this loop many times before most traditional developers finish their first build.

This speed, combined with intentional feedback design, creates a competitive advantage. You are not just building fast — you are building in the right direction because you are letting real users guide you from the start.

Chapter 6: Avoiding Overbuild in the Prototype Stage

One of the most common traps in rapid prototyping is also one of the most dangerous: overbuilding. When your tools are fast, your AI is responsive, and your creativity is firing, it is easy to start adding more and more features "just to see how it looks" or "because it will only take a few more minutes." Before you know it, your simple prototype has ballooned into something complex, harder to maintain, and full of functionality that no one has tested or even asked for.

Overbuilding is a subtle enemy. It does not feel like a mistake in the moment because you are being productive. You are writing code, solving problems, making progress. But in prototyping, progress is not measured by how much you build. It is measured by how quickly you get to a point where you can learn something valuable about your idea. Every extra feature you add before that point is a gamble — it costs time now and may have to be thrown away later.

In vibe coding, the risk is amplified because you can build so much so quickly. When an AI can generate a working module in seconds, you can easily justify adding it "just in case." The danger is that you can end up spending your limited energy building features no one will care about, instead of getting the core of your idea in front of people.

This chapter is about resisting that temptation. You will learn how to spot the warning signs of overbuilding, set guardrails for yourself, and keep your prototypes lean, focused, and easy to evolve.

Why Overbuilding Happens

Overbuilding in prototypes is almost always driven by one of three forces:

1. The Comfort Zone Effect

When a particular kind of feature is easy for you to implement, you gravitate toward it — even if it is not essential to the prototype's purpose. A front-end developer might obsess over styling early on. A back-end developer might over-engineer the database schema. In vibe coding, the AI can amplify this by making your comfort zone features even faster to produce.

2. The "Just in Case" Mindset

You imagine future scenarios in which a certain feature might be needed, so you add it now. The logic is, "It's easy to do now, so why not?" The problem is that the future scenario might never come, and meanwhile you have delayed getting feedback on the parts that matter.

3.Feature Envy

You compare your prototype to mature products and feel compelled to close the gap. If your idea is a task manager, you start thinking, "Well, the big ones all have tagging, reminders, and team sharing — I should add those now." But those products have years of iteration. Your job is to test the core, not replicate their full feature set.

The True Goal of a Prototype

A prototype is not meant to be a smaller version of the final product. It is meant to be a learning tool. Its primary purpose is to validate or invalidate your assumptions, show you what works, and reveal where to go next.

This means that the fewer moving parts you have, the faster you can reach that validation. Every extra feature is another thing that can break, distract users, or produce noise in the feedback you collect.

In vibe coding, the superpower is speed to learning — not speed to feature parity with established products.

The Core Rule: Build for the Question

The simplest way to avoid overbuilding is to keep asking: **"What question is this prototype answering?"**

If a feature does not directly help answer that question, it belongs in the backlog, not in the prototype.

For example:

- If the question is "Will users enjoy generating custom playlists with AI?" then you only need the playlist generation process and a simple way to listen to samples. You do not need user profiles, sharing, or dark mode.
- If the question is "Can we use AI to summarize customer support tickets effectively?" then you only need ticket input, a summarization output, and maybe a quick way to compare to the original ticket. You do not need analytics dashboards or export-to-PDF features yet.

Guardrails to Keep You Lean

Here are practical strategies to prevent overbuilding:

1. Set a Feature Limit

Decide before you start: "This prototype will have three features" or "This will have one primary action and one supporting feature." Anything else goes into the backlog.

2. Timebox the Build

Give yourself a fixed time — one day, one afternoon — to finish the prototype. This forces prioritization and leaves no room for "just one more thing."

3. Backlog Everything

Keep a visible list of every feature idea that comes to mind. This keeps them from distracting you while you build. If the prototype proves its value, you can return to the backlog later.

4. Show It Early

Commit to showing the prototype to at least one person as soon as the main feature works, even if the rest is incomplete. This breaks the perfectionist cycle.

AI-Specific Overbuild Traps

AI makes overbuilding easier in ways you may not expect. For example:

Auto-generated filler: You ask for a small feature, and the AI includes "bonus" features in the code because it assumes you might want them. If you do not actively remove them, you've just increased scope.

Temptation to explore: You can get a working new feature so quickly that adding it feels harmless. Multiply that by three or four features, and suddenly your "quick prototype" is a week-old project.

Over-complex prompts: You try to be thorough in your request, and the AI builds a whole suite of functionality when you only needed one piece.

The solution: be just as disciplined in your prompts as in your coding. Specify what not to build:

"Build a single-page prototype that allows uploading an image and detecting objects in it. Do not add authentication, styling beyond the basic layout, or any other features."

A Real Example: The AI-Powered Interview Coach

Imagine you want to test an AI tool that gives real-time feedback during mock job interviews. Your core question is: Will users find the AI feedback useful and actionable?

You could easily get carried away:

- Adding user profiles so they can save their sessions.
- Building a database to track improvement over time.
- Implementing a scoring system with leaderboards.
- Designing a slick UI with custom themes.

But none of these help answer the core question. The simplest version is:

- A start button that begins recording.
- An AI that transcribes and analyzes answers.
- Feedback displayed in a basic text box.

If users find that feedback valuable, then you can layer on persistence, scoring, and design.

The Feedback Filter

When you show an early prototype, users will suggest features. This is good, but you must filter them. If a suggestion aligns with the question your prototype is trying to answer, consider adding it. If it is about polishing, convenience, or nice-to-have functionality, thank them and add it to the backlog.

Example:

Feedback: "Can it email me the feedback after each session?"Filter: Useful, but not needed to test whether the feedback is valuable in the first place.

Benefits of Staying Lean

Avoiding overbuild has multiple benefits:

- Faster feedback — You can test sooner and more often.
- Clearer results — Feedback focuses on the core feature, not on unrelated extras.
- Easier changes — With fewer moving parts, you can pivot without tearing everything apart.
- Less waste — You avoid building features that will be scrapped if the core idea changes.

Building the Habit

Like any skill, avoiding overbuild takes practice. Start by deliberately making smaller prototypes than you think you need. Use constraints as a creative challenge. Over time, you will learn to feel the point where "enough" is truly enough for a prototype.

One helpful rule is the One-Sentence Test: if you can't describe your prototype's purpose in one sentence without mentioning secondary features, it's too big.

The Vibe Coding Advantage

In traditional coding, avoiding overbuild could feel frustrating because removing scope also meant removing weeks of potential progress. In vibe coding, the opposite is true: you can always add later without much penalty. That freedom should make it easier to stay lean. There is no fear of "missing your chance" to add something — you can add it tomorrow if it turns out to matter.

Avoiding overbuild is not about limiting your creativity. It is about channeling it toward learning as quickly as possible. In rapid AI-powered prototyping, the real measure of success is not how much you can build in one go, but how quickly you can discover the right thing to build. When you master the art of staying lean in the prototype stage, you can move faster, pivot more easily, and arrive at better products with far less waste.

Book 2: Creative Development Workflows

Introduction: Merging Imagination with Engineering

Speed alone is not enough. Rapid prototyping, as we explored in Book **1**, gives you the power to turn ideas into working products in hours or days. But speed without a strong creative approach can produce a stream of functional yet forgettable projects. On the other side of the spectrum, creativity without structure can lead to half-finished experiments, chaotic workflows, and projects that collapse under their own ambition.

The real magic in modern development — especially in vibe coding — happens when creativity and engineering are woven together into a single process. This is not about alternating between "creative time" and "engineering time" like two separate modes. It is about designing a workflow where inspiration and precision feed each other continuously.

AI changes the rules here. In traditional coding, the transition from a creative idea to a solid technical implementation could take days. You might brainstorm a feature, then spend hours figuring out if it's even feasible, then more hours writing the first version. With AI as a collaborator, the feedback loop between idea and execution collapses. You can imagine something, express it in plain language, and see a functioning draft within minutes. That speed removes the friction that used to make creativity and engineering feel like separate worlds.

This book is about mastering that loop. You will learn how to:

- Feed AI prompts that unlock unexpected creative solutions without losing technical quality.
- Maintain flow state for long stretches without burning out.
- Transform AI-generated drafts into production-quality code without slowing to a crawl.
- Reuse and remix existing assets, APIs, and templates in ways that accelerate innovation.
- Treat debugging as a creative problem-solving exercise instead of a tedious chore.

By the end of this section, you will have a set of creative development habits that let you move fast, think big, and build well — all at the same time.

Chapter 7: Turning AI Drafts into Production-Ready Code

The first time you watch an AI generate a complete feature in seconds, it feels like magic. You describe what you want, press enter, and suddenly there it is; code you didn't have to type, logic you didn't have to painstakingly assemble. The rush is real. But if you've ever taken that raw AI output and tried to run it in a production environment without touching it, you know what happens next: edge cases break, performance suffers, and bugs pop up in places you didn't anticipate. AI drafts are powerful starting points, but they're rarely the finish line.

Production-ready code demands more than correctness in the moment. It has to be maintainable, secure, performant, and resilient under conditions the AI wasn't explicitly told to handle. The challenge, and the opportunity, in vibe coding is learning how to keep the speed and momentum of AI generation while adding the layers of refinement that transform a draft into something reliable enough for real-world use.

The Mindset Shift: From Typist to Editor

In traditional coding, you are both the author and the editor of your code. You write the logic yourself, and you review it for quality as you go. In vibe coding, the AI becomes the primary author. Your role shifts toward editor, architect, and quality controller. This doesn't make you less important, it makes your judgment more important.

Think of yourself like the editor of a talented but inexperienced writer. The writer can produce a lot quickly, but they don't always know your audience, your style, or the hidden constraints of your domain. Your job is to guide, prune, rephrase, and restructure until the work meets the standard. The same applies here: the AI's speed is valuable, but it's your engineering sense that ensures the result holds up over time.

Reading AI Output with a Critical Eye

The first step in turning an AI draft into production-ready code is to slow down and read what you've been given. Resist the temptation to copy and paste it straight into your codebase without understanding it. AI code can contain subtle issues; unnecessary complexity, inefficient loops, outdated syntax, or even security vulnerabilities; that aren't obvious until you examine them.

When you read AI-generated code, ask yourself:

- Does this match the architecture and conventions of my project?

- Is it doing extra work that can be simplified?
- Are there hardcoded values or "magic numbers" that should be configurable?
- What happens if the input is invalid, empty, or malicious?
- Is the naming consistent and clear enough for future maintainers to follow?

Even if you're working in a stack you're still learning, you can ask the AI to explain each section of its own output. This helps you spot red flags before they become problems in production.

Iterating for Quality Without Losing Speed

A common fear among developers new to vibe coding is that the review and refinement stage will undo the speed gains of AI generation. That only happens if you try to perfect the code in one pass. A better approach is to work in short, targeted iterations.

You might start with:

"Refactor this function to reduce nesting and improve readability."

Then test. Then follow with:

“Add error handling for network failures and invalid inputs.”

Then test again.

Breaking refinement into focused passes keeps you moving quickly while still improving quality step by step. Each round also gives you a chance to confirm the AI is actually following your style and constraints, not just producing another block of generic code.

The Role of Tests in the Vibe Coding Workflow

One of the fastest ways to lift AI-generated code to production quality is to wrap it in tests as soon as possible. AI can write the tests for you, but the important part is that you run them early and often. Automated tests not only catch issues now, they protect you from regressions later when you or the AI modify the code again.

For example, if the AI writes a data transformation function, you can immediately ask: “Write unit tests for this function with both typical and edge case inputs.” By running those tests right away, you can catch subtle bugs, like incorrect handling of null values or off-by-one errors, before they slip into production.

Security and Performance Checks

Two areas where AI code often needs extra scrutiny are security and performance. AI is good at writing code that "works" in the happy path, but it may not anticipate malicious inputs, high-traffic conditions, or resource constraints unless you explicitly ask it to.

If you're handling user-generated content, for instance, you should always verify that the code sanitizes inputs properly and avoids vulnerabilities like SQL injection or XSS. If the code is processing large datasets, you should check whether the algorithm is efficient enough to handle production loads.

You can speed this process by prompting the AI directly:

"Review this code for security vulnerabilities and suggest fixes."

"Optimize this function for large datasets without changing the output."

The AI can flag many issues instantly, but it's still up to you to validate those suggestions in your environment.

A Real Example: From Draft to Deployment

Let's say you're building a microservice to generate PDF invoices. You ask your AI to create an endpoint that accepts JSON data for the invoice, formats it, and returns a PDF. The AI writes a Flask app that handles the request and uses a library to create the PDF. It works when you test it locally with sample data.

But before shipping it to production, you:

- Review the code and notice it assumes all required fields are present. You add validation and return clear error messages if something is missing.

- Realize the AI's PDF formatting is hardcoded. You refactor it to load a template so you can adjust design without touching code.

- Check performance and see that the PDF generation library blocks the request thread. You switch to an asynchronous call to prevent delays under load.

- Ask the AI to write unit tests for different invoice scenarios, including edge cases like zero-dollar amounts or special characters in customer names.

These refinements take maybe an extra hour, but the result is code that you can trust to handle real customer data without surprises.

Knowing When It's "Ready Enough"

Production readiness is not perfection. The question is not "Is this flawless?" but "Is this reliable enough to serve real users without causing major issues?" AI-assisted workflows can tempt you to polish endlessly because each improvement feels so quick. But part of the skill is knowing when to stop refining and move on.

Set your readiness bar based on the context. For an internal prototype, "ready" might mean the code works reliably with realistic test data. For a public-facing product, "ready" might include passing a suite of automated tests, meeting performance benchmarks, and having error monitoring in place.

The Payoff

When you master the art of editing AI drafts into production-ready code, you get the best of both worlds: the creative speed of AI generation and the robustness of traditional engineering discipline. You stop wasting time on boilerplate and routine implementations, yet you also avoid the trap of shipping brittle code that falls apart in the real world.

And over time, you'll notice the AI starts producing

better drafts from the start, not because it has "learned" in the human sense, but because your prompts and refinements have become sharper. The more you teach it your style, your standards, and your preferred patterns, the more its output will feel like code you would have written yourself.

Chapter 8: Remixing APIs, Templates, and Snippets for Speed

One of the most overlooked skills in vibe coding is knowing how to reuse and recombine existing pieces of work; APIs, templates, and code snippets into something new and useful. Too many developers still default to building everything from scratch, even when an **80**% solution already exists somewhere. The truth is, in the AI-powered era, building from scratch is rarely the fastest path to value. The fastest path is remixing.

Remixing is not laziness. It's leverage. It's the recognition that your real value as a creator is not in retyping the same functions someone else has already perfected, but in connecting the right pieces together in ways that create something unique. A well-chosen template saves hours of design work. An API integration can replace weeks of backend development. A smartly used code snippet can solve a problem before it even has the chance to slow you down.

When you combine remixing with AI, something even more powerful happens: you can blend external resources into your project without having to fully understand every low-level detail before you start.

The AI can bridge the gaps between different libraries, translate patterns from one framework into another, and adapt snippets to fit your specific architecture, often in seconds.

The Mindset of a Remixer

A good remixer is always scanning the landscape for useful building blocks. This is not the same as endlessly scrolling GitHub in the hope that inspiration will strike. It's an intentional awareness that when you encounter a good API, a clean snippet, or a solid template, you should mentally catalog it for later.

For example, you might stumble on a CSS dashboard layout that is responsive, elegant, and lightweight. Even if you don't need it now, you note it down in your personal "remix library." Later, when you need to build a metrics dashboard for a client project, you pull it out and adapt it in minutes with the AI's help.

The key is to see code and tools as modular parts of a larger creative kit. You're not just a programmer. You're a DJ, sampling and layering tracks to create something fresh.

APIs as Creative Shortcuts

APIs are one of the most powerful remix ingredients because they give you access to complex capabilities without having to build them yourself. Want to add AI-driven text summarization? There's an API for that. Need real-time weather data? There's an API for that. Want to send SMS notifications, process payments, analyze images, or run machine learning inference? All available via APIs.

The AI makes integrating these APIs much faster. Instead of manually combing through dense documentation, you can paste the key details into your prompt and say: "Write a function in Node.js that calls the [API name] endpoint to fetch [desired data], handles authentication, and returns the results as JSON." In seconds, you have a working call that would have taken you much longer to piece together from scratch.

The trick is to start thinking of APIs not just as tools for adding features, but as raw material for creativity. Two APIs combined in the right way can produce something completely new. A travel API plus a weather API can produce a "best week to visit" planner. A music API plus a sentiment analysis API can create a "mood-based playlist" generator. The combinations are endless.

Templates as Time Machines

Templates, whether they're for UI, backend

architecture, or entire application scaffolds, are like jumping into a project with a huge head start. Instead of spending hours designing a layout or wiring up basic routing, you start with something that already works and adapt it to your needs.

With AI, the adaptation process is lightning-fast. You can say:

"Take this Bootstrap landing page template and adapt it for a productivity app with these three sections: task list, calendar view, and notes." The AI will not only modify the HTML and CSS for you, but also suggest how to connect it to your backend or integrate it with your chosen APIs.

The mistake most developers make with templates is treating them as fixed artifacts. The power of AI is that it can "melt" a template, re-shape it, and "recast" it into something new without you having to fight with every line manually.

Snippets as the Glue

Code snippets - those small, reusable bits of functionality are the glue that makes remixing possible. They might be a function for formatting dates, a utility for debouncing inputs, or a regex pattern for parsing a string. By themselves, they're small, but when stitched together, they fill in the gaps between bigger components.

The best way to use snippets in a vibe coding workflow is to treat them as modular parts that the AI can drop in and adapt instantly. You can paste a snippet into your prompt and say:

"Integrate this into my current Python script so that it logs all API responses to a file with a timestamp." The AI will handle the integration, adjust variable names, and ensure it fits your current structure.

A Real Remix in Action

Imagine you need to build a "Personal Travel Journal" web app that allows users to log trips, upload photos, and see location-based facts.

- Instead of building a backend from scratch, you use Supabase for authentication and storage.
- Instead of designing a layout from zero, you grab a clean travel blog template from a free template site.
- Instead of writing the location facts feature by hand, you integrate with a public geography API.

On their own, each of these components would take time to integrate. With AI, you can feed each one into your assistant, explain how you want them to interact, and have a functioning prototype in a few hours. What's more, because the AI handles the wiring, you can spend your mental energy on adding small touches that make the app feel personal, like tagging trips with emojis or generating AI-written summaries for each journey.

The Long-Term Value of Remixing

The more you practice remixing, the bigger your "mental inventory" becomes. You start to see possible connections everywhere. You remember that you once used a certain charting library that could fit perfectly into your next analytics project. You recall a template you used months ago that would make a perfect base for a client demo.

And because AI removes so much of the integration friction, you can act on these ideas in the moment. You don't have to shelve them for "when you have time", you can start building right away.

Remixing is a force multiplier. It lets you stand on the shoulders of existing work without being limited by it. Combined with AI, it becomes a way to move at a pace that feels almost unfair. You stop reinventing the wheel and start assembling vehicles that have never been seen before.

Book 3 – Code by Conversation

Introduction: Talking Your Ideas Into Reality

Imagine sitting beside a world-class developer who knows every language, every framework, every library, and every design pattern. You explain what you want in plain language, they nod, and a few moments later they hand you clean, working code. Then you say, "Actually, could we change the layout so the button is above the form?" They do it instantly, with no pushback, no ego, and no billing clock running. This is what coding by conversation feels like when you do it well with AI.

Code by conversation is not just about replacing keystrokes with prompts. It is about shifting the way you think about building software. You are no longer the sole author of every line — you are the director, the architect, and the editor. You shape the vision, guide the execution, and ensure the result meets your standards. The AI does the heavy lifting in between, translating intent into implementation in seconds.

In traditional coding, there has always been a gap

between an idea and a functioning implementation. Bridging that gap required deep technical knowledge, lots of manual work, and often long stretches of frustration when things didn't work the first time. With conversational coding, that gap becomes a dialogue. You can move from "What if we…" to "Here's a working version" in the time it takes to order a coffee.

But this is not just about speed. The real power of conversational coding lies in fluidity. You can:

- Start with a high-level concept and quickly drill into technical specifics without losing context.

- Explore multiple approaches in minutes instead of hours.

- Iterate rapidly, turning rough drafts into refined, production-ready code through a steady back-and-forth.

- Debug collaboratively, treating problem-solving as a guided conversation instead of a lonely slog through stack traces.

In this book, you will learn how to get the most out of that fluidity. We will explore:

Prompting for Precision: How to frame your requests so you get clean, modular, maintainable code on the first try.

Conversational Debugging: How to isolate and fix issues by guiding the AI through targeted, layered questions.

Scaling from Small to Large: How to grow from one-line requests into full, multi-feature applications without the project turning messy.

Style Alignment: How to teach your AI to match your coding conventions so you spend less time rewriting.

Pair Programming with AI: How to use the back-and-forth dynamic to stay productive and creative without burning out.

You will see real examples of how a well-run AI conversation can replace hours of manual work — and how the wrong approach can lead to bloated, brittle code. You will learn how to manage the conversation, not just participate in it.

By the end of this section, you will not just be "using" AI to code. You will be collaborating with it — guiding it like a senior developer guides a junior one, leveraging its speed and breadth of knowledge while steering it with your judgment and creativity. You will be able to talk your ideas into reality, refine them with precision, and scale them into robust applications — all without breaking the conversational flow.

Chapter 9: Conversational Prompting for Clean, Modular Code

One of the defining skills in vibe coding is learning how to talk to your AI in a way that produces not just "working" code, but code that is clean, modular, and easy to maintain. The AI can write in any style you want — structured or messy, reusable or tangled, but it can only match your expectations if you make them explicit. In other words, the quality of your code is often a direct reflection of the quality of your conversation.

When developers first start working with AI, they often treat it like a code vending machine: insert a problem statement, receive a solution. This can work for quick fixes, but if you use AI this way for larger projects, you quickly run into trouble. The output might be functional, but it could be rigid, repetitive, or so tightly coupled that adding new features later becomes painful. The goal in conversational prompting is to shape the AI's output as it's being generated, guiding it toward structures and practices that make sense for your project.

Why "Clean and Modular" Matters

Clean code is not just about making it look nice. It's about reducing cognitive load; for you now, and for whoever works with it later. When functions are clear, names are consistent, and responsibilities are well-defined, you can return to a project months later and still understand it without hours of re-reading.

Modular code is about separation of concerns. Each part of the system does one thing well, and those parts can be reused, replaced, or upgraded independently. When you ask your AI to produce modular code, you're setting yourself up for easier scaling, faster debugging, and smoother collaboration in the future.

AI can absolutely produce this kind of output, but it won't do it by default unless you train it through the way you prompt.

The Flow of a Good Coding Conversation

The most effective AI conversations for clean, modular output feel like a back-and-forth with a seasoned collaborator. You don't just throw out a request and accept whatever comes back. You describe the goal, suggest a structure, review the draft, and refine it in small steps.

For example, imagine you need a feature that generates PDF reports from analytics data. A weak prompt might be: "Write a Python script that takes analytics data and creates a PDF report."

A strong conversational approach might look like this:

"Let's design this so the code is modular and easy to maintain. I want three main parts: one module to format the analytics data, one to generate visual charts, and one to assemble those into a PDF. Start by outlining the function signatures and file structure."

This opening doesn't just tell the AI what to do, it tells it how to organize the work, which sets you up for clean architecture from the start. Once you get the structure, you can move module by module, testing and refining each one before moving on.

Steering the AI Mid-Conversation

One of the biggest advantages of conversational coding is that you can steer the AI without starting over. If you notice the functions are getting too long or the naming is inconsistent, you can pause and say: "Refactor this so each function is under **20** lines and the variable names are more descriptive."

The AI will adjust while keeping the overall logic intact. This is where modular thinking pays off: when the code is already broken into logical pieces, changes like this are far easier for the AI to handle without introducing new bugs.

Building a Shared Vocabulary with Your AI

If you work with the same AI model often, you can develop a shorthand for your preferred style. Over time, you might not have to say "make it modular" every time. You can simply refer to "our standard structure" or "the pattern we used in the last project." The AI will draw on your earlier conversations to match that style.

This shared vocabulary can include things like:

- Your preferred naming conventions.
- How you handle configuration and constants.
- Whether you prioritize performance or readability when there's a trade-off.

The more consistent you are in your language, the more consistent the AI will be in its output.

A Real Example: Cleaning as You Go

Let's say you're building a REST API in Express.js for

a small inventory management tool. You give the AI a quick prompt and it generates a working version in a single file. Technically, it's fine. The API responds to requests, data gets stored, and you can demo it. But a quick look at the code tells you everything is crammed together: the routes are defined inline, the business logic is tangled in with request handling, and database queries are scattered in between. It's the kind of code that works today but will turn into a headache tomorrow.

Instead of accepting that as the "finished" version, you stop and guide the AI: "Separate the routes, controllers, and database queries into different files. Each controller should only handle request validation and call the appropriate service. Use async/await for all database operations, and make sure errors are handled consistently."

The AI instantly reorganizes the code into a clean structure:

Routes handle only the paths and the mapping to controllers.

Controllers handle input validation and orchestrate calls to the service layer.

Service functions encapsulate the database queries, keeping them isolated from request logic.

Now, each piece of the system has a clear, singular purpose. You can see where every responsibility lives, and making changes in one part won't ripple unpredictably through the rest. If you decide to switch from MongoDB to PostgreSQL, all your changes happen in the service files. If you need to add authentication, you can focus entirely on the controller layer without touching the core business logic.

The best part? You didn't have to spend an hour manually refactoring. You just asked the AI, mid-build, to restructure the output to your standards. You kept your momentum while improving your architecture in real time.

Why This Matters in Vibe Coding

Vibe coding thrives on speed and flow, but speed without structure is just a shortcut to technical debt. It's tempting to accept the first working draft and keep adding features on top, but every unstructured decision you make early will cost you more later.

Conversational prompting for clean, modular code changes that equation. It lets you move quickly without locking in bad habits or fragile architecture. Instead of thinking of cleanup as a separate, dreaded phase, you make it part of the ongoing conversation with your AI. It becomes a natural rhythm: generate, review, refine.

Once you're comfortable with this style, you can tackle more complex systems without fear that they'll collapse under their own weight when you scale. You'll know that each piece is built to stand alone, to be replaced, extended, or optimized without tearing the whole thing apart.

Clean, modular code is your insurance policy. And in an AI-driven workflow, it's an insurance policy you can enforce instantly, not by digging into every line yourself, but by telling your AI exactly how you want it done.

Chapter 10: Debugging as a Dialogue, Not a Struggle

Debugging has always been the part of programming that tests patience the most. It's the grind behind the glamour, the quiet hours spent chasing a bug through a maze of files, logs, and stack traces until you finally track down the culprit. For many developers, debugging is where focus dies. You start out determined to fix a small issue, and two hours later you're deep in unrelated code, wondering why your function is still returning null.

With vibe coding, debugging changes. The AI becomes your partner in the investigation, reducing the amount of mental energy you need to spend on grunt work and letting you focus on decision-making. Instead of you alone sifting through every possibility, you can ask targeted questions, share snippets of code or error messages, and get instant suggestions that are directly relevant to the problem at hand.

From Frustration to Conversation

In the traditional debugging workflow, the first step is often to guess where the problem might be, then start combing through the code. If you're lucky, the issue is obvious - a missing semicolon, a typo in a variable name, a function called before it's defined. But in real-world applications, bugs are rarely that kind. They're subtle. They only show up under certain conditions. They're buried in dependencies you didn't write. And the process of finding them can be slow and demoralizing.

When you work with AI, the process feels more like talking through the problem with a colleague. You can paste the error message and the relevant code into your prompt and ask: "Here's the function and the stack trace. Walk me through the likely causes of this error."

The AI doesn't just spit out a single guess. It can give you a short list of possibilities, explain why each one might be responsible, and even suggest quick experiments to confirm or rule them out. This moves you out of the purely reactive mode, chasing one hunch after another, and into a guided process where each step is deliberate.

Keeping the Conversation Tight

One of the most important habits in AI-assisted debugging is resisting the urge to throw your entire codebase into the conversation at once. It's a natural impulse, when something's broken, you want to give the AI "all the information" so it can find the problem quickly. But ironically, too much context can have the opposite effect. When the AI is flooded with irrelevant details, its answers tend to become vague, unfocused, or based on assumptions that don't target the real issue.

Instead, you want to feed it the smallest, most relevant slice of code and context that still reproduces the problem. Think of it like giving a doctor only the symptoms that matter, not your entire medical history. If you hand over **5,000** lines of unrelated code to explain a bug that lives in **50** lines, you're forcing the AI to wade through noise before it can see the signal.

This doesn't mean hiding information. It means delivering it in layers. Start with the smallest reproducible example, the single function or component where the error shows up. Let the AI respond, and if it needs more, it will ask for it. That's your cue to widen the view, showing the parent module, then the related data flow, and eventually the way that module interacts with the rest of the system.

This back-and-forth is where the AI's conversational nature shines. Each exchange builds a clearer, more accurate picture of the bug's environment, without wasting time on irrelevant code paths. You're effectively zooming out one level at a time, like adjusting a camera lens until the whole scene comes into focus.

The result is sharper, faster debugging sessions where the AI stays focused on the actual problem rather than trying to guess its way through a flood of unnecessary context. You get targeted solutions, and the process feels collaborative rather than chaotic.

Using AI to Build Your Debugging Map

When debugging manually, you often have to create a mental map of how the bug travels through the system; which functions call which, what data is being passed, where state changes occur. AI can generate that map for you in minutes. You can paste in the relevant parts of the code and say:

"Trace the flow of data from this API endpoint through the service layer to the database write operation. Summarize it step by step."

Now you have a clear, written path to follow. You can annotate it, run tests at each stage, and pinpoint exactly where things stop behaving as expected.

Turning AI Into Your Test Partner

One of the most powerful ways to debug with AI is to treat it not just as a code generator, but as a collaborative test engineer. Instead of manually writing every test case yourself, you can lean on the AI to design and run targeted tests that quickly confirm or eliminate your theories about where the bug lives.

Once you have a suspicion about a problem area, don't just dive in and start rewriting code. Pause and frame your suspicion as a test request. For example, you might say: "Write a unit test that isolates this function and tries edge case inputs."

The AI can generate a ready-to-run test file, complete with assertions, setup, and teardown logic if needed. This isn't just a time-saver. It's also a way to ensure you're testing systematically rather than chasing fixes based on gut feeling alone.

If your bug is elusive, maybe it only appears under specific load conditions, data shapes, or timing scenarios; you can ask the AI to simulate those conditions. You might say:

"Write a performance test that calls this API endpoint **1,000** times with varying payload sizes and logs any failures."

Or: "Simulate a data stream where **10**% of the messages are malformed JSON and confirm the parser handles them gracefully."

These kinds of controlled, repeatable tests are where the AI really earns its keep. They speed up the confirm-or-deny loop - that critical part of debugging where you're trying to determine whether a theory about the bug is valid. Without AI, that loop can take hours because you're building each test manually. With AI, you can generate and run a new test in minutes, adjust it based on the results, and move on to the next hypothesis without losing momentum.

Over time, you'll notice another benefit: the AI starts producing test cases that not only catch the current bug but also guard against similar issues in the future. In other words, each debugging session can leave your project a little more resilient than it was before. That's how you turn AI from a reactive helper into a proactive quality partner.

A Real Example: Debugging by Conversation

Let's say you're working on a Next.js app and your dynamic routing suddenly breaks in production, even though it worked fine locally. Without AI, you might spend hours reading through build logs, re-checking route configuration, and searching for similar issues online.

With AI, the process could go like this:

You paste the routing configuration and the relevant part of the error message into your prompt: "In production, this route throws a **404** error even though it works locally. Here's the config. What are the most likely causes?"

The AI responds with a handful of plausible reasons, such as differences in base path configuration, case sensitivity in file names on different operating systems, or missing static build paths. It explains each one briefly. You try the easiest fix first; adjusting the base path, and redeploy. If that doesn't solve it, you move to the next suggestion. Within an hour, you've resolved the issue without touching half the codebase.

Debugging as a Learning Process

Debugging with AI isn't just faster, it's a better learning tool. Every time you hit a problem, you can ask the AI to explain not just the fix, but why the problem happened in the first place and how to prevent it in the future. Over time, these lessons accumulate. You start to recognize patterns before they cause trouble, and your instinct for spotting potential pitfalls improves.

This is a big shift from the old debugging cycle, where you might fix a bug without fully understanding it just to move on. Now, each fix is also a mini-lesson, which makes you sharper with every project.

Letting AI Handle the Repetitive Parts

Some parts of debugging are just plain tedious. Searching logs for matching entries. Comparing two versions of a file line by line. Checking every function call for a specific parameter across an entire codebase. Manually doing these tasks not only eats up time but also chips away at your focus. These are the perfect kinds of jobs to hand off to your AI.

The process is simple: you give the AI the raw input, whether it's a log file, a set of file diffs, or a directory of code, and the criteria for what you're looking for. The AI can then surface the relevant details in seconds while you stay focused on interpreting the results and deciding what to do next.

In a standard one-off interaction, you might paste in an error log and say:

"Scan this log for any entries related to **NullReferenceException** and summarize the timestamps and affected modules."

Or: "Compare these two versions of **apiClient.js** and highlight only the changes to function parameters."

But the real leap comes when you move into what

ChatGPT calls Agent Mode, where the AI is allowed to take a series of steps on its own, iterating toward a g[illegible]each inst[illegible]ader dire[illegible]

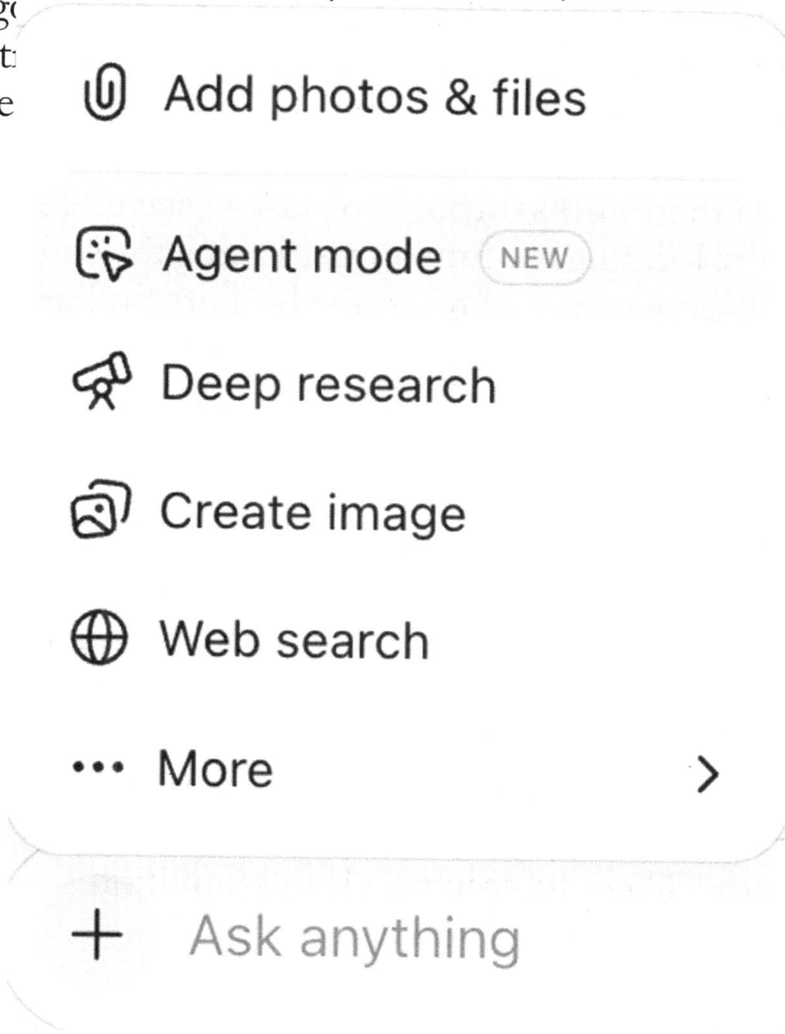

"Trace the flow of this bug from the error log to the responsible function, check for similar patterns elsewhere in the codebase, and suggest a fix."

The AI then works through the task in stages; reading the logs, locating the function, scanning the codebase for related issues, and returning both its findings and a draft solution. You're still in control, reviewing its steps and approving changes, but you're no longer acting as the middleman for every subtask.

By letting the AI, especially in Agent Mode, take care of the repetitive and mechanical parts of debugging, you protect your mental energy for the high-value work: diagnosing the deeper cause, validating the fix, and thinking about the long-term stability of the system. You keep the process in a steady rhythm instead of letting it spiral into fatigue, and you're far more likely to reach a clean resolution without burning out halfway through.

Debugging will never be the most glamorous part of coding, but with the right approach, it doesn't have to be the most frustrating. When you treat it as a dialogue rather than a solo grind, you stay calmer, move faster, and come away from each bug not just with a fix, but with a deeper understanding of your system. And that's the real win, because every solved bug becomes another tool in your belt, ready to make the next challenge a little easier.

Chapter 11: Scaling Up From One-Liner Requests to Multi-Feature Apps

One of the most exciting moments in vibe coding is realizing that the same AI that can help you write a small, one-off script can also help you build a complete, multi-feature application. The leap from "one-liner prompt" to "full product" is both thrilling and a little dangerous. It's thrilling because you can produce something substantial faster than ever before. It's dangerous because, without the right structure, speed can lead to chaos; spaghetti code, overlapping logic, or a tangled web of dependencies that's hard to untangle later.

The key is to treat scaling as a gradual process, not an explosion. A one-liner prompt can give you a working starting point, but scaling it into a robust, multi-feature system requires intention. You need to maintain clarity, keep features modular, and preserve the ability to adapt as the app grows. In other words, the challenge is not just adding more, it's adding more without losing control.

The Temptation to "Just Keep Adding"

It starts innocently enough. You prompt your AI to "build a simple to-do list app" and in minutes you have one running in your browser. You test it, it works, and you feel a rush of possibility. So you ask for a feature: "Add the ability to categorize tasks." The AI updates the code. Then you ask for another: "Add due dates." Another: "Add drag-and-drop sorting." Within an afternoon, you've bolted on half a dozen features, each one stacked on top of the last.

It still works, but somewhere along the way the clean simplicity you started with has been replaced by complexity. Functions have grown longer. State management feels messy. Adding new features starts to cause breakages in unrelated parts of the app. The initial speed has turned into slowdown.

The problem isn't the AI. The problem is that without a clear scaling plan, you've been building vertically, piling features on top of each other, instead of horizontally, where each new feature stands on its own and connects cleanly to the rest.

Scaling with a Layered Approach

A better way to grow an AI-assisted build is to work

in layers. Think of your initial one-liner request as the "core layer", the simplest version of the app that delivers the main function. Before you add more, take a step back and design the next layer: a feature that can be added without breaking the core, ideally as a self-contained module.

For example, if your core is the to-do list, you might design the "categories" feature so that it lives in its own module, with its own logic for creating, storing, and displaying categories. The to-do list simply calls into this module when it needs category data, instead of having category logic scattered throughout. The AI can help enforce this structure if you ask for it: "Add a categories feature to this app, but keep it in its own module and have the to-do list import from it rather than mixing the logic."

This approach slows you down slightly in the moment, but it makes the rest of the scaling process far faster and safer.

Growing the Conversation Alongside the Code

When your app is small, your prompts can be short. But as the app grows, your prompts should grow too. Instead of just saying, "Add a new feature," you need to give the AI the context it needs to integrate it well. That means describing:

- What the new feature should do.
- How it should connect to existing features.
- Any constraints or architectura patterns you want to preserve.

You don't have to paste the entire codebase every time, but you should share the relevant files and remind the AI of the structure you've been following. This helps it maintain consistency instead of introducing design drift.

Testing as You Scale

The more features you add, the more likely it is that something will break when you change the code. This is why scaling and testing go hand in hand. AI can generate the tests for you, but you need to run them frequently, ideally after each feature is added.

Think of tests as your "safety net" for scaling. Without them, every new feature is a leap into uncertainty. With them, you can add and modify features with confidence, knowing that if something breaks, you'll catch it right away.

Example: From One-Liner to Full Tool

Imagine starting with this one-liner:

"Build a Python script that takes a text file of customer feedback and outputs the top three recurring themes."

You run it, and it works. That's your core. Now you want to turn it into a full customer feedback dashboard. You scale horizontally:

- Add a file upload interface (in its own module).
- Add a visualization feature to show themes over time (in its own module).
- Add sentiment analysis for each theme (in its own module).

At each step, you tell the AI to integrate the new feature without merging its logic into unrelated parts of the app. When you look at the code months later, each feature is still cleanly separated, and you can update or replace them individually.

Knowing When to Stop Scaling

Just because AI makes it easy to add features doesn't mean you should add them endlessly. The best apps, especially prototypes, have a clear scope. If your goal is to validate an idea or show a demo, scaling beyond the essential features can actually slow you down.

Set a "feature ceiling" for each version of the app. Once you hit that ceiling, shift focus from adding new things to refining what you have. A polished, reliable app with a small feature set is far more impressive than a bloated app that only works half the time.

The Real Win

Scaling with AI is about more than speed, it's about keeping your future self (and your collaborators) happy. If you structure your growth, maintain modularity, and keep the conversation with your AI focused, you can expand from a one-liner to a multi-feature app without losing the joy and momentum that vibe coding brings. And that's the sweet spot: moving fast, but building something that lasts.

Chapter 12: Teaching Your AI to Match Your Standards

One of the hidden superpowers in vibe coding is that the AI can adapt to you. It's not just a neutral code generator. It can learn your preferences, mimic your style, and anticipate the kinds of structures you want. The closer the AI gets to producing output that matches your standards on the first try, the faster and smoother your workflow becomes. But this doesn't happen automatically. You have to train it.

"Training" here doesn't mean fine-tuning in the machine learning sense. It means developing a shared language and set of expectations through repeated interaction. The more clearly and consistently you communicate what "good" looks like to you, the more the AI will default to producing that kind of output. Think of it like onboarding a new team member. You wouldn't expect them to guess your conventions and preferences on day one, but over time, they'd pick them up if you guided them well.

Why Standards Matter in Vibe Coding

Without clear standards, AI output tends to be generic. It works, but it might not match your architecture, your naming conventions, or your preferred way of handling certain scenarios. You end up spending time reformatting, renaming, and refactoring; which slowly eats away at the speed advantage of vibe coding.

Standards solve this problem by creating a "style contract" between you and the AI. They define the invisible rules that shape your code:

- How functions should be named and organized.
- How errors should be handled.
- Whether comments should be verbose or minimal.
- How configuration and environment variables should be managed.

Once the AI understands these rules, it can produce code that drops cleanly into your project without a lot of rework.

Making Your Standards Explicit

The first step in teaching your AI is to state your preferences clearly. This means going beyond "write clean code" and actually defining what clean means to you. For example:

- "Use snake_case for variables and functions, PascalCase for classes."
- "Limit functions to **25** lines or less, breaking them into smaller helpers if needed."
- "Always include docstrings that describe the parameters and return values."
- "Catch and log all exceptions, but let the error propagate unless otherwise specified."

You can give these as a single set of instructions at the start of a session or paste them into your prompt whenever you start a new project.

Using Examples as Style Guides

One of the most effective ways to align the AI with your standards is to show it examples of your own code. You can paste a file or snippet and say: "This is the style I want you to follow for this project."

The AI will analyze things like indentation, naming patterns, comment style, and even how you structure imports. From that point on, you can refer back to "the earlier style example" and it will attempt to match it. Over multiple projects, you can build a small "style reference" that you reuse as part of your prompting toolkit.

Correcting and Reinforcing

When the AI outputs something that doesn't match your standards, don't just fix it silently, tell it what to change. For example: "This is close, but remember to use snake_case for variables. Please refactor to match."

This feedback loop reinforces the rules and makes it more likely the AI will follow them next time. If you notice the same kind of mismatch happening repeatedly, you can add a permanent note to your standard instructions.

Teaching Project-Specific Standards

Sometimes your standards aren't global. They're specific to a project. Maybe one app uses a particular folder structure, or a certain API client must be wrapped in a custom service class. In those cases, you can onboard the AI just like you would a new developer joining the project.

You might say: "In this project, all API calls must go

through the **apiClient.js** module. Do not call fetch directly. When you add a new endpoint, define it in **apiClient.js** and import it where needed."

By stating the rule once, you save yourself from having to repeatedly correct direct fetch calls throughout the code.

Capturing Your Standards for Reuse

If you notice yourself giving the same kinds of instructions to the AI again and again, "use snake_case for variables," "limit functions to **25** lines," "write docstrings for every public function", that's your signal to stop repeating yourself and start codifying your preferences. The fastest way to do this is to capture them in a reusable format.

This doesn't have to be complicated. A simple text file, a note in your preferred knowledge management tool, or a page in your project wiki is enough to start. The goal is to create a short, portable document that contains your standard prompts, coding style rules, and architectural preferences. Whenever you start a new project, you paste them into the very first conversation with your AI. That initial injection sets the tone for everything that follows.

Over time, this collection becomes your personal AI coding profile, a living document that evolves as you refine your approach. Early on, it might be only a handful of rules. Months later, it could include detailed guidelines for naming conventions, error handling, testing practices, security measures, and preferred library usage. The beauty of having it written down is that you don't have to rely on memory or reinvent your instructions for every project.

This profile also becomes incredibly valuable when switching between AI tools. Each model has its own quirks, but your standards are portable. Instead of starting from scratch every time you try a new assistant, you can import your profile on day one and get consistent output faster. You're not just retraining the AI each session, you're training your workflow to be predictable, repeatable, and aligned with your own quality bar.

Some developers even take this further by keeping multiple profiles; one for personal side projects, another for enterprise-level builds, another for rapid prototypes where polish is less important. Each profile reflects a different "mode" of working, and all are just a paste away from being active in a conversation.

Once you get into the habit, you'll notice a major shift: you spend less time correcting AI output for style or structure and more time building. The AI starts producing results that feel like your code, not generic code, right from the start, because you've taken the time to make your standards part of the conversation before any code is written.

A Real Example: Applying Standards in Action

Let's say you're building a small SaaS admin dashboard with your AI assistant. You paste in your standards:

- Use TypeScript with strict mode enabled.
- Organize components into /components, /pages, and /utils.
- Use functional components with hooks; no class components.
- Prefix all async functions with "fetch" or "load" to make their purpose clear.

The AI starts building with these rules in mind. You notice one file breaks the naming convention for async functions, so you ask it to refactor. Within seconds, the file is updated and consistent. By the time you've added a dozen components, the entire project feels unified, even though much of it was generated in short bursts of conversation.

The Long-Term Payoff

The more you teach your AI, the less you have to fix. Your projects become easier to maintain, your commits feel cleaner, and you spend more time adding value instead of reformatting output. Over time, you'll also notice that your prompts become shorter because the AI already understands your baseline.

The goal is not to make the AI perfect. It will still need adjustments, but to get it producing work that feels like it could have been written by you or your team. When you reach that point, you're no longer just using the AI. You've made it part of your development culture.

Book 4 – Low-Code Empowerment

Introduction: Building Big Without the Heavy Code

For decades, the path to building functional software was narrow and steep. You needed to know a full programming language, understand databases, handle deployment, and stitch together every moving part yourself. That level of control was powerful, but it was also a barrier. Many good ideas never made it past the "wouldn't it be cool if…" stage because the gap between inspiration and implementation was simply too wide for one person to cross quickly.

Low-code platforms changed that. By providing pre-built components, visual builders, and ready-to-use integrations, they lowered the entry barrier. You no longer had to write every line of code yourself to get a product into users' hands. You could drag, drop, and connect your way to something functional in a fraction of the time.

But low-code on its own has limits. Anyone who has worked in these platforms for long knows the moment when you hit the wall — that frustrating point where you want to do something just slightly outside what the built-in features allow. In the past, that was where your momentum stopped or where you had to hand the project off to a full-stack developer to "finish it properly."

With AI in the picture, that wall is no longer a dead end. Low-code tools become launch pads, not cages. You can still use their speed and simplicity for the bulk of your work, but now you can also:

- Generate custom features on demand when the platform's defaults are not enough.
- Integrate with APIs the tool does not natively support.
- Refactor and optimize the generated code to improve performance or add flexibility.
- Design unique user experiences that do not look like a default template everyone else is using.

This is low-code empowerment — the fusion of rapid visual building with AI-driven customization. It lets you move faster than traditional coding without inheriting the rigid limits of pure low-code.

In this part of the book, you will learn how to:

- Pick low-code tools that give you speed now and freedom later.
- Connect multiple systems, APIs, and databases without falling into "syntax hell."

- Upgrade low-code builds with custom AI-generated code that makes them stand out.
- Decide when it is time to move from low-code to a full-code environment — and how to make that move gracefully.
- Maintain performance, security, and scalability even in a low-code context.

The goal here is not to convince you to use low-code for everything. It is to show you how to extract the maximum speed from these platforms while avoiding the long-term traps. By the time you finish this section, you will be able to start and finish serious projects in low-code without worrying about hitting an unfixable limit — because you will know exactly how to break through any barrier the platform throws at you.

Chapter 13: Choosing Low-Code Tools That Don't Limit You Later

Low-code tools have been around for years, but the rise of AI has changed the way we look at them. In the past, many developers dismissed low-code platforms as "training wheels" for people who didn't know how to code. They were seen as quick ways to throw together a prototype, but not something you'd trust for a serious product.

That perception is shifting fast. AI can now plug directly into low-code workflows, making them far more powerful and flexible than before. You can use AI to generate custom components, integrate APIs that the platform doesn't natively support, and even write scripts to extend the platform's capabilities. Suddenly, low-code is not just for non-developers. It's a legitimate way for experienced developers to move faster without giving up control.

The catch is that not all low-code tools are created equal. Some will give you room to grow; others will box you in so tightly that scaling becomes a rebuild. The challenge is knowing which is which before you invest your time and data.

The Three Kinds of Low-Code Platforms

When you strip away the marketing, most low-code tools fall into one of three categories:

1. Closed Ecosystem Builders

These platforms give you everything you need - database, UI builder, automation tools, but keep you inside their walled garden. They're fast to start with but can be slow to evolve if your needs go beyond what's built in. Examples include Glide, AppSheet, and some tiers of Airtable.

2. Extensible Platforms

These give you a visual builder and basic backend tools but also allow you to drop in custom code, integrate any API, and export or migrate your data easily. They strike a balance between speed and flexibility. Examples include Bubble, Retool, and WeWeb.

3. Low-Code Layers on Top of Full-Code Frameworks

These tools accelerate development in a real programming environment rather than replacing it. They give you prebuilt components, drag-and-drop interfaces, or scaffolding commands that save time but still let you dive into raw code when needed. Examples include RedwoodJS with scaffolds, Supabase Studio, and some Next.js admin generators.

The safest long-term bet is usually the second or third category, something that lets you stay fast but doesn't lock the door when you need to get under the hood.

What "Doesn't Limit You Later" Actually Means

When we say a low-code tool "won't limit you later," we mean it protects you in three ways:

1. **Data Freedom** — You can export your data in a usable format, not just a proprietary blob.

2. **Code Access** — You can add or modify code beyond what the visual builder allows.

3. **Integration Reach** — You can connect to outside APIs, services, or databases without begging the platform vendor for permission.

If a tool fails on any of these three, it might still be fine for a one-off prototype, but you should think carefully before using it for anything that might grow.

The AI Advantage in Low-Code

AI turns low-code into something far more potent by erasing many of its historical weaknesses. In the past, if a platform didn't support a specific API or had a quirky limitation in how it handled data, you'd have to wait for the vendor to update it or learn a workaround yourself. Now you can simply ask your AI to generate the missing piece; a connector script, a data transformation, a custom UI widget, and drop it into the platform's code injection feature.

This means you can start with a low-code platform's speed and still overcome most of its technical limits without switching platforms prematurely.

How to Evaluate a Low-Code Tool for Long-Term Use

When testing a low-code platform, don't just look at how quickly you can make something work. Look at how easily you can break out of the defaults. Some good tests to run early:

- Can you connect it to an external database and still use its UI builder?
- Can you trigger workflows based on events from outside the platform?
-

- Can you export your project in a format that's still readable and usable outside the platform?
- How does it handle authentication, and can you replace it with your own system if needed?

The answers to these questions tell you whether the tool is a fast lane or a cul-de-sac.

A Real Example: Picking the Right Tool for a Growing Product

Imagine you want to launch a simple customer onboarding portal. In a pure prototype mindset, you could throw it together in something like Glide in a weekend. But you know there's a good chance it will need custom dashboards, API integrations with your CRM, and possibly real-time data sync.

You decide to start in Bubble instead. It's still low-code, so you can build most of it visually, but it lets you drop in custom JavaScript and integrate with external APIs right away. Six months later, when your onboarding flow needs a live analytics dashboard, you use your AI assistant to generate the custom charting module and plug it in without rethinking the whole system.

Because you chose a platform that could stretch with you, scaling didn't mean starting over.

Choosing for Prototypes vs. Products

When you're in pure prototype mode, your only real priorities are speed and simplicity. The mission is to get something in front of users, or even just in front of yourself, as quickly as possible so you can see if the idea has legs. In that context, it doesn't matter if the database is proprietary, the integrations are limited, or the platform has quirks you'd never tolerate in production. You're not here to future-proof; you're here to future-discover.

For example, if you want to test a concept for a personalized meal planner, you might spin it up in Glide or AppSheet over a weekend. You can hack together a basic UI, pull in a handful of recipes, and let a few friends try it. You're not worrying about whether the platform's database can handle **100,000** recipes or whether you can swap out the UI framework later. At this stage, speed is the only metric that matters, because the point of a prototype is to answer the question: "Is this worth pursuing?"

But the moment you're building something you expect to last, even in a small way, the calculation changes. Now you have to think about exit options. What happens if the platform's pricing model changes in six months? What if you need an integration that the vendor doesn't support? What if your user base doubles and the performance starts to degrade?

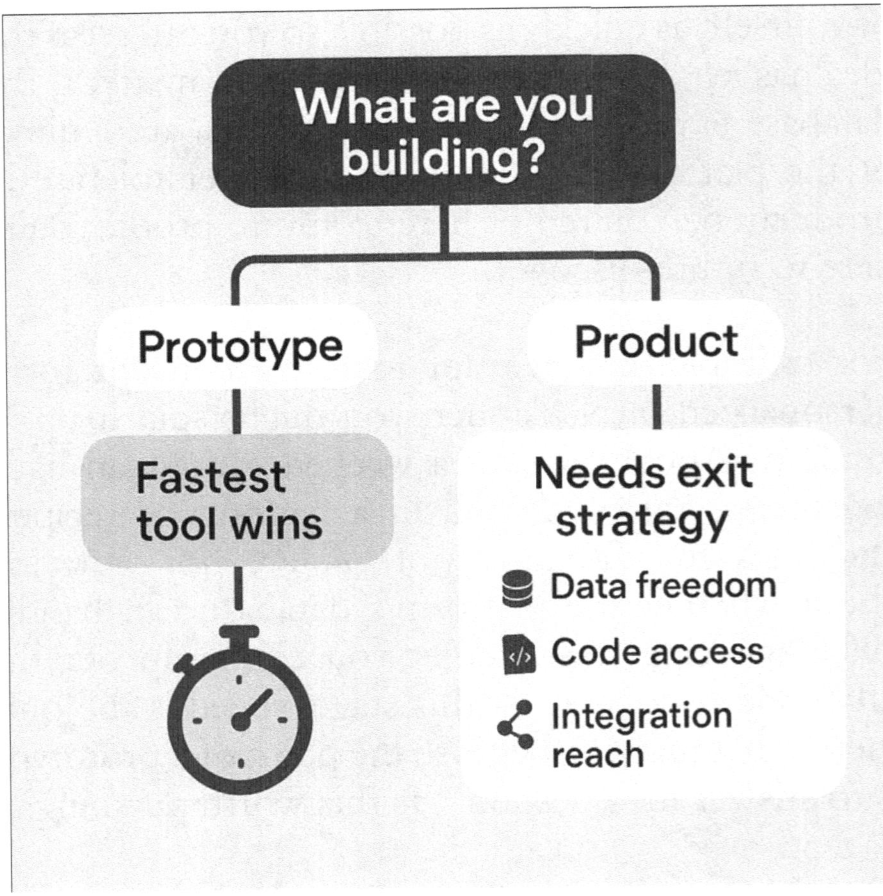
What are you building?
Prototype
Product
Fastest tool wins
Needs exit strategy
Data freedom
Code access
Integration reach

A low-code platform that's perfect for a prototype might be a trap for a product. The wrong choice can force a painful rebuild when you hit the limits. The best low-code choice for a lasting product is the one that lets you grow inside it for as long as possible and gives you a clear path out when you've outgrown it. That means:

- Your data can be exported in a usable format without a fight.
- You can inject custom code when you need more control.
- You're not locked into a single hosting option.

If your product is something you'll be maintaining for months or years, pick a tool with an escape hatch. It might take a little longer to set up than the fastest pure-prototype platform, but you'll save yourself the nightmare of tearing everything down and starting over when you inevitably need more than the platform can give.

In vibe coding, the skill isn't just building quickly. It's building quickly in a way that doesn't paint you into a corner. The difference between a throwaway prototype and a long-lived product is not just in the code, but in the choices you make about the foundation it sits on.

The real art of choosing low-code tools in vibe coding is balancing the thrill of speed with the wisdom of future-proofing. The wrong tool can make you feel like a genius in week one and a prisoner in month six. The right one will let you keep moving quickly today while keeping every door open tomorrow.

Chapter 14: Connecting Data, APIs, and Automation Without Syntax Hell

In traditional coding, connecting multiple systems together; a database here, an API there, an automation flow somewhere else; can feel like assembling a piece of furniture without the instructions. You have all the parts, but the fit isn't obvious, the screws are slightly different sizes, and you spend hours just making things align. For many developers, this "glue work" is the least glamorous part of the job. It's necessary, but it's slow, and it often involves wrestling with syntax, authentication quirks, and unpredictable edge cases.

Low-code platforms already make this easier by providing connectors, visual data mappers, and prebuilt automation steps. But these tools still have limits. The moment you need something slightly outside the built-in options, like a custom API request with unusual headers or a data transformation the visual tools don't support, you risk falling back into syntax-heavy coding that breaks the flow.

This is where AI changes everything. In a vibe coding workflow, the AI doesn't just write code for you, it becomes your bridge between incompatible systems, translating formats, generating glue logic, and smoothing over the friction points that usually slow integration to a crawl.

From Integration Anxiety to Integration Confidence

In the old way of working, starting a multi-system integration could feel daunting. You'd open three different documentation tabs, scan through endless parameter lists, and hope you didn't miss the one sentence buried in the API docs that explained the authentication scheme.

With AI, you can bypass much of this. You can paste the relevant snippet from the API docs into your prompt and say:

"Write the request code to get data from this API using my API key stored in an environment variable."

Or, if you're in a low-code tool with a custom code block, you can drop in exactly what the AI gives you and be running in minutes.

You don't need to hold the entire syntax in your head. The AI does that for you. Your focus stays on what you want connected and why, not on memorizing the exact format of a cURL request or how to handle pagination for a specific API.

Connecting Data Sources Without the Pain

Data is the lifeblood of most modern apps, but it's rarely sitting in one neat, ready-to-use place. You might have user data in a CRM, transaction data in a separate database, and analytics data in a third-party service.

A low-code platform might give you connectors for each one, but making them work together still requires mapping fields, cleaning formats, and sometimes transforming structures so they match.

In a vibe coding workflow, you can have AI handle the heavy lifting here too. For example:

- You export a sample dataset from each source.
- You paste them into your prompt with an explanation: "I need these merged so that customer IDs match and dates are all in YYYY-MM-DD format."
- The AI gives you a transformation script that you can run inside your low-code platform's scripting block or in a separate ETL step.

This turns what could be hours of trial-and-error mapping into a few guided interactions.

Automation Without the Syntax Trap

Automation is another area where syntax hell likes to creep in. Even in visual workflow tools like Zapier, Make, or n8n, you eventually hit a step where the built-in modules don't quite do what you want. That's when you're prompted to write a "code" step, and if you're not fluent in the exact language it requires, your momentum stalls.

In vibe coding, you can ask your AI to generate exactly the snippet you need for that step. Instead of figuring out the quirks of JavaScript in Zapier's Code by Zap step, you tell the AI: "Write a JavaScript snippet for Zapier that takes an array of objects and returns only those where the 'status' field is 'active'."

You paste the result, test it, and move on. No deep syntax study, no guessing at the right property names, just quick execution.

A Real Example: Multi-System Harmony in Hours

Imagine you're building a client dashboard in a low-code platform like Retool. You need to:

1. Pull customer account data from Salesforce.

2. Merge it with recent invoice data from Stripe.

3. Trigger a Slack notification for your account managers when an account goes overdue.

Here's how it plays out in a vibe coding flow:

- You grab the API docs for Salesforce and Stripe, paste the relevant parts into your AI prompt, and ask it to generate the requests using the platform's API connector format.

- You export sample data from both APIs and have the AI write a merge function that lines up the customer IDs, cleans the date formats, and calculates how many days overdue each account is.

- You feed that merged data into a Slack automation step, again having the AI generate the exact payload Slack's API expects.

By lunch, you've stitched together three systems into one automated dashboard, all without spending hours deep in syntax or documentation.

Thinking Like a Connector, Not Just a Builder

One of the biggest mindset shifts in modern development; especially in a low-code, AI-assisted world is learning to see yourself less as a solitary builder hammering out every piece of code from scratch and more as an orchestrator of components. Those components might be APIs, databases, automation workflows, third-party SaaS tools, or even open-source libraries. Your role is not to reinvent each piece, but to make them work together in a way that creates something new and valuable.

Think of it like composing music. You don't need to build every instrument yourself. Instead, you select the right instruments, arrange them so they complement each other, and ensure the timing and flow create a coherent whole. In vibe coding, AI is like having a session musician who can instantly adapt to any instrument you pick, it can help you "translate" between systems that were never designed to speak the same language.

For example, you might connect a CRM's API to a shipping provider's API, feed that combined data into a reporting dashboard, and then automate follow-up emails when certain triggers occur. None of those components were built with your exact use case in mind, but by acting as a connector, you create a flow of information and action that solves a unique problem.

In this approach, your value comes less from memorizing syntax for a hundred different systems and more from knowing which systems to connect, in what order, and for what purpose. It's about pattern recognition; understanding the flow of data and functionality well enough to identify the best linkages. AI accelerates this because you can describe what you want to achieve in plain language, and it will generate the translation layers, API calls, and glue code to make it happen.

This is a much more strategic role than traditional coding. Instead of measuring productivity by how many lines of code you've written, you measure it by how effectively you've combined existing resources into something that delivers real value. It's also far more scalable. Once you think like a connector, you can tackle increasingly complex projects without increasing your workload proportionally, because you're leveraging systems, not building every piece from scratch.

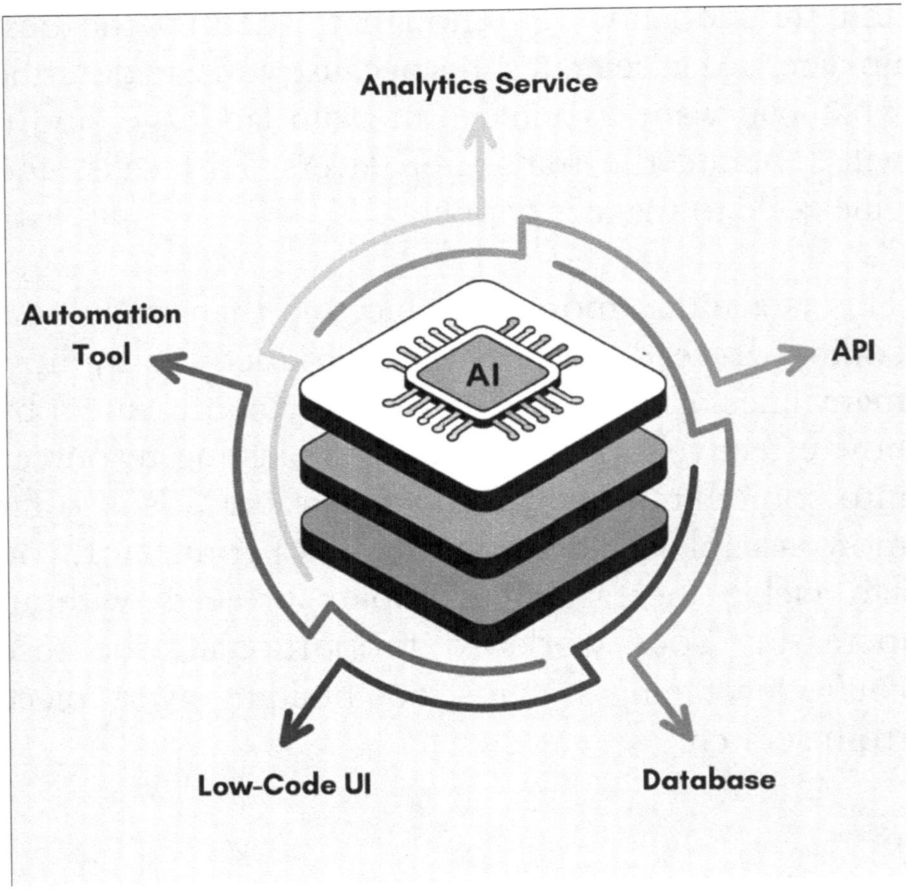
Analytics Service
Automation
Tool
AI
API
Low-Code UI
Database

In vibe coding, the most impactful developers are often the best connectors. They know how to assemble powerful ecosystems of tools, services, and automations; and they know how to use AI as the conductor that keeps the whole system in sync.

By removing syntax hell from the equation, AI turns integration into something you can approach with confidence and even creativity. You're no longer limited by what you personally remember or can type from scratch. You can connect, merge, and automate across platforms with the same ease that you add a single new feature to an app. And once you start working this way, the idea of grinding through integrations the old-fashioned way will feel as outdated as manually writing every line of HTML without a text editor.

Chapter 15: Upgrading Low-Code Projects with Custom AI-Generated Features

Low-code platforms are fantastic at getting you from zero to a working product quickly. You can drag-and-drop your interface, connect to a few data sources, set up automation steps, and have something ready to demo in hours. But if you've ever used a low-code platform for anything beyond a prototype, you know there's a catch: sooner or later, you hit a wall.

That wall might be a missing feature, a limitation in the platform's logic builder, or a design element you can't quite replicate with the built-in tools. It might be an integration that only works halfway or a performance bottleneck the vendor hasn't solved.

In the past, this was where many low-code projects stalled. You could either live with the limitation, try to hack around it with clumsy workarounds, or rebuild in full code. But now, with AI in your toolkit, there's a fourth option: extend the platform yourself, even if you've never written that kind of code before.

Why AI Makes Extensions Possible for Everyone

AI erases much of the intimidation factor that comes with extending low-code platforms. You no longer have to be an expert in the platform's scripting language or its API to build custom features. You can describe what you want, paste in the platform's documentation, and have the AI generate exactly the code you need to slot in.

This means you can:

- Add features that the vendor hasn't built yet.
- Replace clunky built-in modules with more efficient versions.
- Create unique, brand-specific interactions that make your app stand out.
- Integrate with services the platform doesn't natively support.

The best part is that you're not just getting a "hack", you're getting maintainable, documented code that you can reuse and adapt.

Spotting Upgrade Opportunities

The first step in upgrading a low-code project is knowing where to focus. Look for points where the platform's default options either slow you down or fail to deliver exactly what you want. Common examples include:

- **Data processing** that's too slow or too limited in the built-in tools.

UI elements that can't be styled or behave the way you need.

Integrations that lack key parameters or event triggers.

Automation logic that stops short of your ideal workflow.

Once you've identified one of these friction points, frame it as a specific request for your AI: "I need a function that takes the raw data from this API, cleans it according to these rules, and returns it in the format my low-code platform expects."

Case Study: Adding AI to a Non-AI Platform

Let's say you've built a customer support dashboard in Bubble. The platform gives you a clean UI builder and a database, but it doesn't have built-in natural language processing to summarize long customer messages.

Instead of waiting for Bubble to add this as a plugin, you can:

1. Use your AI assistant to generate a connector to an external API (like OpenAI or Anthropic).

2. Have the AI write the JavaScript needed to send the customer's message to the model and receive the summary.

3. Integrate that script into Bubble's workflow so that every new message automatically gets summarized before it's stored.

Now your dashboard has an AI-powered feature that isn't even on the platform's roadmap.

Making Upgrades Sustainable

When you add custom AI-generated features to a low-code project, you want to avoid creating fragile code that breaks the moment the platform updates. You can reduce this risk by:

- Keeping all custom code in clearly labeled, isolated modules.
- Documenting what each custom feature does and how it's connected.
- Asking the AI to comment the code in plain language so future maintainers can understand it.

This way, even if you need to migrate later, you'll be able to lift your custom work out cleanly and reuse it elsewhere.

The Competitive Edge

One of the most overlooked (and underestimated) benefits of upgrading low-code builds with AI is the competitive edge it gives you in a crowded market. Low-code platforms are designed for accessibility, which is their greatest strength but also their greatest weakness in terms of differentiation. When hundreds of people use the same platform with the same built-in templates and preconfigured modules, their apps inevitably start to look, feel, and behave the same. The UI patterns are familiar, the workflows predictable, the capabilities capped at whatever the vendor has packaged for everyone.

That sameness is where your opportunity lives. By injecting AI-assisted custom features, you break free from the gravitational pull of "default." You can build interactions, automation, and intelligence that simply can't be replicated by drag-and-drop alone. A client logs into your Bubble app and notices it has real-time AI-powered recommendations that adjust based on their behavior, something no standard plugin offers. A stakeholder sees your Retool dashboard and realizes it's proactively summarizing data trends in natural language rather than waiting for them to dig through charts. These touches make your product feel alive, dynamic, and tailor-made.

This isn't just about "fancy features" for the sake of novelty. It's about creating genuine value in ways that your competitors can't match without your skill set. While they're bound by the walls of the low-code platform, you're able to extend it in any direction you choose. You can tap into APIs the platform doesn't natively support, create complex data transformations that happen invisibly behind the scenes, and design user experiences that feel handcrafted instead of mass-produced.

The ability to combine low-code speed with AI-powered custom extensions is one of the most future-proof skills you can develop as a builder. It means you can operate at startup speed without inheriting the limitations of the platform you started on. You get the best of both worlds: the rapid prototyping, the reduced maintenance overhead, and the accessibility of low-code, paired with the limitless creativity and technical reach of custom development.

And because AI can write, refactor, and adapt these custom features for you in real time, you don't need a large dev team to compete at this level. You can deliver results that feel like they came from a team of ten when, in reality, it's just you, your low-code tool, and your AI co-pilot working in perfect sync. That's not just a competitive edge, that's a competitive moat.

When you learn how to see every platform limitation as an opportunity to extend rather than a reason to stop, your low-code work stops feeling like a compromise and starts feeling like creative freedom. With AI as your silent coding partner, there's almost nothing you can't add, and that makes you a different kind of developer entirely.

Chapter 16: Knowing When to Switch from Low-Code to Full-Code

Low-code platforms are like turbo-charged launch pads. They get you off the ground faster than traditional coding, often with less effort and fewer people. But no launch pad is meant to be a permanent orbit; at some point, the platform that once gave you speed can start to slow you down.

The moment you hit that point, you face a decision: keep pushing within the platform's constraints, or make the leap to full-code. The wrong decision can cost you months of time and energy. Stay too long, and you may end up with a brittle, overcomplicated setup. Switch too soon, and you may spend weeks rebuilding features that didn't need to be rewritten yet.

The good news is that if you're deliberate about it, the transition doesn't have to be a painful restart. In vibe coding, the same AI that helped you build your low-code app can help you migrate it to a full-code environment in a controlled, even exciting way.

Why Switching Happens

Most low-code to full-code transitions happen for one or more of these reasons:

- **Performance Limits** — The app starts to lag with larger datasets, higher user counts, or more complex workflows.

- **Feature Gaps** — You need custom functionality the platform can't (or won't) support.

Integration Barriers — You hit limits in API connectivity or real-time data handling.

- **Cost Curve** — As your usage grows, the monthly platform bill becomes higher than hosting your own custom build.

- **Ownership Concerns** — You want full control over your code, data, and hosting environment.

These are not failures of low-code. They're natural growing pains; signs that your prototype or MVP has outgrown its first home.

Reading the Signs Early

The best time to start planning your switch is before you absolutely have to. If you wait until the platform is actively blocking your progress, the transition becomes urgent and stressful.

Some early warning signs include:

- You're writing more and more custom scripts to bypass platform limitations.
- You've maxed out your database or API call quotas several months in a row.
- You're paying extra for features that should be core to your workflow.
- Your team spends more time finding workarounds than building new value.

When these signs start appearing, it's time to start preparing your migration plan, even if you don't act on it right away.

Planning a Smooth Migration

Switching to full-code doesn't have to mean throwing everything away. In fact, the best migrations treat the low-code app as a living prototype that informs the new build.

Start by:

1. **Documenting the current system** — Export workflows, data models, and any custom scripts you've built.

2. **Separating the essentials from the extras** — Decide which features must be in the first full-code version and which can wait.

3. **Choosing your target stack** — Use AI to help you pick a combination of frameworks, databases, and hosting solutions that match your performance and scalability needs.

4. **Designing in modules** — Build the new app in components so you can bring features online one at a time instead of all at once.

This approach keeps the transition incremental and reduces the risk of a big, fragile launch.

Using AI as Your Migration Partner

AI can dramatically shorten the migration process. You can paste your low-code workflows or exported schema into your AI assistant and ask it to:

- Generate equivalent full-code modules.
- Suggest optimizations for performance and maintainability.
- Scaffold an entire project in your chosen framework.

If your low-code app has a lot of custom scripts, AI can also refactor them to fit the conventions of your new stack, saving you the manual labor of translating each function yourself.

Example: Migrating Without Losing Momentum

Imagine you've built a small booking platform for local events in Bubble. It works well for the first year, but now you're hosting bigger events, the user base has grown, and load times are becoming noticeable. Plus, you want to add real-time seat selection, which the platform doesn't support well.

You decide to migrate to a Next.js app with a PostgreSQL backend hosted on Supabase. You start by exporting your Bubble database schema, then feed it into your AI with the request: "Generate a Supabase schema and migration script from this structure."

Next, you export your workflows and have the AI translate them into serverless functions. You rebuild the UI module by module, testing each against the live Bubble version until you're confident enough to cut over. Within a few weeks, you have a faster, more flexible app without losing the features your users rely on.

Knowing When Not to Switch

Sometimes the smartest move is to stay in low-code longer. If your app is still comfortably meeting user needs, performing well, and generating revenue, there's no rush to migrate. A well-chosen low-code platform can handle production workloads for years; especially when extended with AI-generated custom features.

The key is to make migration a strategic choice, not a reaction to frustration.

Switching from low-code to full-code isn't about abandoning one for the other, it's about recognizing that each has its season. Low-code is your launch runway. Full-code is your long-distance flight. And with AI guiding both phases, the transition can be less of a jolt and more of a natural evolution in your build's life cycle.

Book 5 – Next-Gen Explorer Mindset

Introduction: Staying Ahead in the Age of AI

The tools, languages, and platforms you are using today will not be the same ones dominating the field in a few years — in some cases, even a few months. In the age of AI, the rate of change in software development has accelerated beyond anything the industry has seen before. New frameworks emerge weekly. Existing platforms evolve so quickly that the features you relied on last quarter might be obsolete by the next. AI models themselves improve in capabilities at a pace that would have seemed impossible a decade ago.

For many developers, this pace feels overwhelming. The temptation is to dig in and cling to the tools you know, to build a comfort zone and stay inside it. The problem is that comfort zones in tech shrink over time. The longer you stay still, the more the industry moves past you. Skills that were cutting-edge yesterday become average today and irrelevant tomorrow.

The next generation of successful builders — whether they are solo makers, startup founders, or team leads — will not be defined by a single stack, a single platform, or a single way of working. They will be defined by adaptability. They will be explorers.

- An explorer mindset means you:
- Approach unfamiliar technologies without hesitation.
- Treat learning as a continuous part of building, not a separate phase.
- Experiment with new tools and workflows regularly.
- Evaluate trends quickly to decide which are worth adopting.
- Can pivot when market or technical conditions shift.

AI is your compass in this new landscape. It removes much of the friction that once made exploration risky or slow. You can try a new programming language in an afternoon, integrate with an unfamiliar API in an hour, or spin up a proof-of-concept with a brand-new framework on the same day you hear about it. The cost of trying new things has never been lower.

But exploration for its own sake is not enough. The real skill is strategic exploration — knowing how to scout new technologies efficiently, test their value quickly, and integrate them into your toolkit when they prove useful. It is also about knowing when to walk away from something that is not worth the investment, no matter how shiny or hyped it might be.

In this part of the book, you will learn how to:

- Build skills in unfamiliar technologies without fear or long delays.

- Scout and test new AI development tools in under an hour.

- Turn small experiments into revenue-generating products.

- Keep your ethical standards and long-term sustainability in mind as you innovate.

- Develop a repeatable process for staying ahead of emerging trends without getting distracted by every passing novelty.

By the time you finish this section, you will not just be keeping up with the AI-driven evolution of software development — you will be thriving in it. You will have the confidence to step into any new environment, learn what you need, and build something meaningful before others have even decided whether to try. You will stop being a passive consumer of tools and trends and become an active navigator of the future.

Chapter 17: Building Skills in Unfamiliar Tech Without Fear

Every developer has a comfort zone. It's the stack you know by heart, the tools you can work with while half-asleep, the patterns you've internalized so deeply that you barely have to think about them. Comfort zones feel safe, and safety has its place. But they can also quietly become cages. The moment a project demands something outside that zone, fear can creep in.

That fear often takes the form of hesitation. I've never used that framework before. I don't know that language. I don't have time to learn it right now. It's easy to default to the tools you know, even if they're not the best fit for the job. Over time, this habit can shrink your world, making you less adaptable and less competitive in a fast-moving field.

The core of the next-gen explorer mindset is learning to treat unfamiliar technology not as a threat, but as an invitation. The unknown isn't a barrier. It's a fresh path to explore. And with AI at your side, that path is no longer steep and rocky. It's walkable from day one.

Why Fear Disappears in a Vibe Coding Workflow

Before AI, learning a new technology often meant a long, lonely ramp-up: tutorials, docs, trial-and-error builds, and many hours of confusion before you could produce anything worth showing. That time investment was the real barrier.

In vibe coding, you skip much of that ramp. The AI can act as a real-time tutor, translator, and co-builder. You don't need to memorize syntax or fully understand the internals before you start making progress. Instead, you can say: "I've never used Vue before. Set up a basic app with a single component that fetches and displays JSON data from this endpoint."

Within minutes, you have something running. As you build, you can ask the AI to explain what each part does, suggest best practices, and help you add features. You're learning in context, and that makes the knowledge stick faster.

Building Confidence Through Micro-Wins

One of the fastest ways to kill fear is to rack up small, visible wins. In an unfamiliar tech stack, those wins might be as simple as:

- Getting a "Hello World" app running.
- Connecting to an API successfully.
- Rendering your first dynamic list.

In vibe coding, you can accelerate these wins by asking the AI for the smallest working example of what you're trying to do. That first working piece becomes the anchor you build from. Each layer you add (styling, interactivity, persistence) is another confidence boost.

Fear thrives in abstraction. The moment you can see something real on the screen, fear starts to fade.

Asking Better Questions

When you're working in a new technology, you might not even know what you don't know. This is where AI's conversational nature shines. You can start with vague, high-level questions and gradually drill down.

For example:

- "How does state management work in React?"
- "Show me a simple counter component in React with state management."
- "Refactor this counter to use a custom hook for incrementing."

Each question builds on the last, and because the AI remembers your earlier steps, it can keep explanations relevant to your current context.

Combining Exploration With Real Projects

Learning sticks best when you have a reason to use it. Instead of learning a technology in isolation, pick a small real-world project (even something trivial), and build it in the new stack. This turns your learning into problem-solving rather than passive study.

For example, if you want to explore Go, you might build a simple CLI tool for renaming files in bulk. If you want to try Svelte, you could build a minimalist weather widget. The AI can handle the rough spots while you focus on understanding the parts that matter most.

Accepting Imperfection

When you're new to a tech stack, you won't produce flawless code on day one. That's fine. In fact, it's expected. The key is to get comfortable with building before you fully understand. In vibe coding, you can rely on AI to fill in gaps and catch mistakes, which means you can work at speed without the fear of "messing it up." You'll clean and refine as you go.

The perfectionist mindset is the enemy of exploration. If you wait until you feel "ready," you'll never start.

Example: Fearless First Steps

Say you've never touched Flutter, but you need to make a quick mobile prototype. Instead of disappearing into documentation for a week, you tell your AI: "Build a basic Flutter app with a single screen that has a button. When pressed, it fetches a random joke from this API and displays it."

Within minutes, you're running it in an emulator. You can then ask:

"Explain how the state is managed here."

"Refactor to separate the API logic into its own file."

By the end of the day, you've learned the basics of Flutter's widget system and state management, all while producing something tangible.

Fear of unfamiliar technology comes from distance. The longer you stand outside looking in, the bigger the gap seems. In the next-gen explorer mindset, you close that gap immediately. You start building on day one. You ask for help the moment you hit friction. You let curiosity lead, and you let AI handle the steep parts of the climb.

Over time, this habit makes you dangerous in the best possible way: no matter what tool, language, or framework the future throws at you, you'll know you can step in, get moving, and make it yours.

Chapter 18: Scouting and Testing New AI Dev Tools Efficiently

The pace at which new AI-powered developer tools are appearing is unlike anything the software industry has seen before. What feels like a breakthrough on Monday becomes baseline by Friday, and by the following week something new is already in private beta promising to replace it. This constant churn can be both exciting and exhausting. You hear about a new code generation tool, a sleek framework for building AI-powered apps, a cutting-edge API that promises to save hours of work. The temptation is always there to drop whatever you are doing and try it immediately.

But here's the truth: not every tool deserves your attention. Many will be overhyped but underdelivering. Some will vanish within months, leaving you scrambling to replace them. A few will quietly become indispensable — but you will only recognize them if you have a way to evaluate them quickly and systematically. Without a process, you risk spending hours or even days on exploration without gaining anything you can actually use in your real work. In vibe coding, where your time and focus are your greatest assets, that is too high a price.

This is why scouting and testing must be intentional. It is not enough to know that new tools exist. You need a structured way to move from raw curiosity to a clear verdict in the shortest possible time, and you need to capture your findings so you never repeat the same evaluation unnecessarily.

From Curiosity Overload to Curated Discovery

If you have ever clicked through to a new tool's homepage, skimmed its marketing copy, glanced at the documentation, tinkered with it for a bit, and then closed the tab with a vague sense that it was "cool" but maybe not for you — you have experienced curiosity overload. You walked away without a concrete answer to the question: Should I use this?

In vibe coding, we can do better than that. The goal is to take that initial spark of interest and, in a short, focused burst of exploration, arrive at a practical verdict: Does this fit my workflow right now? Should I adopt it, revisit it later, or ignore it? This requires you to be deliberate about what "good" looks like for you. A tool that is game-changing for another developer might be irrelevant for you if it doesn't align with your style, your needs, or your projects.

Step 1: Define Your Fit Criteria

Before you even sign up for an account or install anything, decide what would make this tool worth adopting. The more specific your definition of "fit," the less likely you are to waste time. For some developers, speed is the only thing that matters — if the tool does not make you faster than your current setup, it fails immediately. For others, flexibility might be the deciding factor — a tool that handles multiple project types might be more valuable than one that is locked to a single workflow.

Think about how it integrates with what you already use. Does it work alongside your existing stack, or would it force you to change your whole setup? Consider stability as well — a flashy demo is meaningless if the tool is buggy or unreliable under real conditions. Finally, be realistic about the learning curve. If you cannot be productive with it in under an hour, is it worth the investment?

What matters here is that these criteria are personal. Your fit profile will not be the same as another developer's. This is not about chasing someone else's "top **10** tools" list. It is about knowing exactly what earns a spot in your toolkit.

Step 2: Run a One-Hour Build Test

Once you have your criteria, give the tool a real test. Set aside exactly one hour and try to build something tangible. Not a "hello world" example from the docs — something that reflects the kind of work you actually do. The point is to simulate a real task so you can see how the tool performs in the context that matters to you.

For example, if you are evaluating an AI-powered SQL query builder, you might connect it to a test database, write a simple SELECT query, add a JOIN, perform an aggregation, and then generate a quick visualization of the results. If all of that happens smoothly within an hour, the tool has passed a meaningful test. If you find yourself stuck halfway through, wrestling with vague documentation or discovering that key features are missing, then you already know it is not ready for your real projects.

This build test is where you cut through hype. Marketing can make any tool look good in theory. Only working with it on a task you care about reveals whether it truly fits.

Step 3: Compare Against Your Current Workflow

A new tool does not have to be perfect to be worth adopting. It simply needs to improve on what you already have in a significant way. After your one-hour build test, take a moment to compare the experience to your current setup. Did you get more done? Did you feel less mental strain? Did the tool enable something you could not easily do before?

If the answer to even one of these is "yes," the tool may deserve a place in your workflow — or at least a deeper look. If the answer is "no" across the board, then it is safe to move on without regret. This is not about loyalty to the tools you already use; it is about making deliberate, measurable improvements.

Step 4: Capture the Verdict

One of the easiest mistakes to make as a tool scout is failing to record your conclusions. You try something, form an opinion, and move on — only to find yourself evaluating the same tool months later because you forgot why you passed on it.

The solution is to keep a simple "Tool Journal." It could be a text file, a Notion page, or even a spreadsheet. Record the tool's name and URL, the date you tested it, a short summary of what you tried, and your verdict. The verdict can be as simple as "Adopt," "Revisit later," or "Drop." Over time, this log becomes your personal map of the AI development landscape. It prevents wasted time, and it reveals patterns about the kinds of tools you consistently find valuable.

A Real Example: A Fast Pass on a Hyped Tool

Suppose you hear about an AI-powered CSS generator that promises to create perfect, responsive styles from plain English descriptions. You are intrigued but skeptical. You decide to run a one-hour build test with the challenge of creating a responsive card layout for a portfolio site, complete with hover effects.

You feed the description into the tool, and it produces CSS instantly. When you integrate it, it works — but you notice it is bloated with unused styles. You spend **20** minutes cleaning it up. The total build time is still shorter than writing the CSS yourself, but not by much. You record in your Tool Journal: "Useful for quick prototyping, but not for production without heavy cleanup. Keep for design sprints only." That verdict now sits in your log, ready to save you from retesting the same thing later.

The AI Factor in Scouting

AI changes the scouting process in two important ways. First, it can help you learn a new tool faster. Instead of trawling through forums or digging through incomplete documentation, you can ask the AI directly how to accomplish specific tasks inside the tool. This collapses the learning curve dramatically.

Second, AI can help you stress test the tool more thoroughly. It can suggest edge cases or advanced features to try — things you might not have thought of on your own in a first pass. This means your one-hour build test is not just about scratching the surface; you can push the tool toward its limits right away and see how it handles under pressure.

The Payoff of a Structured Process

Scouting new AI development tools is no longer about passively collecting bookmarks or making vague promises to "check them out someday." With an explorer mindset, you have a defined process: evaluate quickly, decide firmly, and document your results. You stay current without drowning in noise, and you keep your toolkit sharp without losing weeks to experimentation that goes nowhere.

The result is that when the right tool appears, you will not hesitate. You will recognize it, know exactly how to test it, and be ready to put it to work immediately — not months later when the opportunity has passed.

Chapter 19: Turning Solo Experiments into Revenue-Generating Products

One of the most empowering shifts in vibe coding is realizing that you don't need a big team, a huge budget, or months of work to create something people will pay for. In the old model of software development, going from idea to revenue often required a startup-level effort: long development cycles, multiple roles, marketing resources, and significant funding. Now, with AI and low-code tools in your corner, you can create, launch, and monetize a useful product in weeks, sometimes even days.

The barrier between "experiment" and "product" is thinner than it has ever been. The skill is learning how to cross that barrier quickly and deliberately.

The Nature of AI-Era Experiments

Most experiments in vibe coding start small: a micro-project you built to learn a new tool, a quick automation you made for your own workflow, or a proof-of-concept feature you coded for fun. Alone, each one might feel too simple to sell. But with a few targeted improvements- polish, packaging, and positioning; many of them can be turned into something marketable.

The trick is spotting experiments with external value. If your small build solves a problem that other people also have, or if it creates a delightful experience others would want, it's a candidate for monetization.

Step 1: Spotting Monetizable Potential

When evaluating an experiment for product potential, ask:

- Does this solve a problem others have frequently enough to pay for?
- Could someone use this without my direct involvement?
- Is the result better, faster, or cheaper than what's already out there?
- Could I explain its value to a stranger in under **30** seconds?

An automation that turns meeting transcripts into clean, shareable notes? That has obvious value. A dashboard that visualizes sales trends from Shopify data? Also valuable. A whimsical tool that turns selfies into pixel art avatars? Maybe niche, but still monetizable if positioned well.

Step 2: Adding the Product Layer

Turning an experiment into a product usually means adding three layers:

Usability — Make it easy for someone else to use without setup headaches.

Reliability — Ensure it works consistently under typical conditions.

Presentation — Give it a clear name, a simple interface, and a basic web presence.

For example, if you built a quick AI-powered contract summarizer for yourself, turning it into a product might involve:

- Wrapping it in a web app so users can upload their own files.
- Handling errors gracefully when a file is too large or unsupported.

Adding a landing page that explains the tool in plain language with a "Try Now" button. You don't have to make it perfect. The goal is to get it to a "minimum lovable product" level, where people see the value and can try it immediately.

Step 3: Picking a Revenue Model

With small, AI-assisted products, you have a lot of flexibility in how to monetize:

- **One-time purchase** for simple tools that don't require ongoing hosting or maintenance.
- **Subscription** for tools that provide continuous value or evolving features.
- **Usage-based pricing** for API-heavy products where costs scale with use.
- **Freemium** with basic features free and advanced features behind a paywall.

The key is matching the revenue model to the product's nature. A personal productivity tool might do well as a one-time purchase. A data dashboard with live updates probably fits a subscription model.

Step 4: Launch Fast, Learn Faster

The worst thing you can do after building something is let it sit unpublished while you overthink. Launching doesn't have to mean a global marketing campaign. It can be as simple as posting in a relevant forum, sharing on social media, or emailing people in your network who might benefit.

AI makes even the launch process faster: it can help you write a landing page, create demo videos, draft outreach emails, and design simple marketing assets. The goal is to get real users interacting with your product as soon as possible so you can learn what to improve.

Step 5: Iterating with Purpose

Once you have users, your feedback loop becomes the most valuable tool in your process. Ask:

- Which features are people actually using?
- What do they ask for repeatedly?
- Where do they get stuck or confused?

Instead of guessing what to add next, let your users guide you. In vibe coding, you can ship small improvements rapidly, sometimes same-day, which means you can respond to feedback almost in real time.

Example: From Weekend Project to Revenue

A developer builds a "Podcast Clipper" tool over a weekend. It takes an audio file, finds interesting moments using an AI model, and exports them as short, shareable clips. Initially, it's just a personal tool to speed up editing.

But after showing it to friends, they realize podcasters everywhere could use it. The developer:

- Wraps it in a simple upload interface on a low-code platform.
- Adds payment handling with Stripe.
- Writes a short landing page explaining the benefits.

Within a month, they're making recurring revenue from a handful of subscribers, and they keep iterating based on real usage.

The Mindset Shift

The big shift here is seeing every build not just as a learning exercise, but as a potential seed for a business. Not every experiment will grow into a product. and that's fine. But by thinking this way, you start to notice more opportunities.

With AI accelerating both the build and the polish stages, the gap between "cool idea" and "real product" has never been smaller. The only question is how many times you're willing to cross it.

Chapter 20: Ethical Guardrails and Sustainable AI Development

The speed and power of vibe coding can feel intoxicating. You can build complex applications in days, experiment with new ideas in hours, and automate entire workflows before lunch. But with that speed comes responsibility. In traditional development, long build cycles and multiple layers of review acted as natural brakes; slowing you down, yes, but also giving time to spot potential harms. In an AI-accelerated workflow, those brakes are gone. If you're not intentional, you can ship something before you've truly thought through its impact.

Sustainable AI development is about building in a way that's not only technically sound but also ethically responsible and long-term viable. It means asking questions before you launch, setting boundaries on what your systems can do, and keeping in mind the people who will be affected, both users and non-users.

Why Ethics Feels Different in AI Development

In traditional software, the main risks were usually tied to technical flaws; bugs that break a feature, downtime that frustrates users, or poor performance that slows everything down. These were important problems, but they tended to be visible, measurable, and relatively contained. Fixing them was often a matter of patching code, upgrading infrastructure, or tightening QA.

In AI-assisted development, the risks extend far beyond the code's internal mechanics. Here, the danger isn't just that something won't work. It's that it will work too well in ways you didn't intend, or in ways that harm people. An AI-powered product can unintentionally amplify existing social biases, surface inappropriate or offensive outputs, mishandle or leak sensitive data, or be repurposed for harmful use cases you never envisioned.

The speed of vibe coding magnifies these risks. When you can create and deploy in hours, the distance between "I have an idea" and "This is affecting real people" is incredibly short. There's little time for the traditional layers of review and deliberation that might have caught potential issues in a slower development cycle. A proof-of-concept you built for fun can, within days, be in the hands of hundreds or thousands of users, for better or worse.

This acceleration changes the moral equation. In traditional builds, you might have weeks or months to identify, debate, and mitigate possible harms before launch. In AI builds, ethical thinking can't be bolted on afterward, by the time you get to "after," the product may already be out in the world. That's why it has to be part of the build process itself, as integral as writing tests or designing the user interface.

Ethics in AI development isn't just about avoiding "bad press" or regulatory trouble. It's about actively designing for safety, transparency, and fairness from the first line of code (or the first AI prompt). The decisions you make at the very start, which data sources to use, how to handle user input, what limits to put on automation; can determine whether your product becomes a trusted tool or a cautionary tale.

Building with Guardrails from Day One

Guardrails aren't about limiting creativity. They're about channeling it in a way that avoids preventable harm. In vibe coding, you can set these guardrails early and still move quickly.

Some key areas to consider:

- **Data Privacy** — Are you handling user data responsibly? Are you collecting only what you need? Is it stored securely?

- **Transparency** — Do users know when they're interacting with AI? Do they understand how the system works at a high level?

- **Bias Awareness** — Could your AI outputs be skewed in ways that disadvantage certain groups? Do you have checks in place to spot this?

- **User Control** — Can users override or correct AI-generated content?

Even in a rapid build, you can address these by asking your AI: "Review this design for privacy and ethical risks and suggest improvements."

Sustainability Beyond the Launch

Ethics is one side of sustainable development; longevity is the other. Many AI-assisted builds are launched quickly but are not designed to last. Sustainability means thinking about maintainability, cost, and scalability right from the start.

Ask yourself:

- Will the hosting and API costs still make sense if the user base grows **10**x?
- Is the architecture modular enough to evolve without a full rebuild?
- Can someone else maintain this if I step away?

In vibe coding, it's easy to think short-term because the build is so fast. But fast code can still be durable if you design it that way.

Example: Responsible Scaling in Action

Suppose you build an AI-powered resume analysis tool. It takes an uploaded CV, parses it, and gives tailored improvement suggestions. The first version works well, and you get positive feedback. But before scaling, you consider the ethical and sustainable angles:

You add a privacy policy that clearly explains how uploads are handled and deleted.

You run tests to check if the AI is giving biased advice based on gendered language or certain education backgrounds, and you adjust prompts accordingly.

You move from a basic hosting plan to one that can handle spikes in traffic without crashing.

This extra step doesn't slow you much, but it ensures your product can grow without creating legal or reputational risks.

The Explorer's Responsibility

Part of being a next-gen explorer is realizing that your reach grows as your tools get faster. You can affect more people, more quickly, than builders in the past. That's a privilege, and it's also a responsibility.

Ethical guardrails aren't about avoiding risk entirely — that's impossible. They're about reducing unnecessary harm, being transparent about limitations, and keeping the human impact in mind. Sustainable thinking is about making sure the things you build today can stand tomorrow without constant patching or damage control.

When you combine speed, creativity, and responsibility, you don't just build quickly. You build well. You create things that can survive in the wild, earn trust, and make a positive impact. And that is the true mark of mastery in vibe coding: not just the ability to make something fast, but to make something worth keeping.